THE A

M000206964

Alastair Bonnett is Professor of Social Geography at Newcastle University. His previous books include *Off the Map*, *Beyond the Map*, *New Views: The World Mapped Like Never Before*, *What is Geography?* and *How to Argue*. He lives in Newcastle.

'Extraordinary... Bonnett writes with an acerbic charm... A fascinating and intelligent book. It brings geography to life in a way that felt-tip drawings of Dutch polders never could.' *Sunday Times*

'Fascinating... Man-made territories provide the most interesting moments in Alastair Bonnett's tour of our planet's many islands.' *Daily Mail*

'A beguiling, fact-filled account of the world's headlong dash to build artificial islands. Via a mixture of extensive desk research and short field visits, Bonnett invites readers to journey with him from military-orientated "Frankenstein Islands" in the South China Sea to gigantic windfarms anchored to the bottom of the North Sea.' *Times Literary Supplement*

'A knowledgeable world tour of different types of islands, much enhanced by self-deprecating accounts of his own often shoestring visits... Bonnett has written a most readable and sympathetic account of the various guises islands can take around the world and rightly points out the ecological consequences of human building projects.' James Hamilton-Paterson, *Literary Review*

'A great primer on the concept of islands in the modern age. . . Engagingly written.' *Library Journal*

'In *The Age of Islands*, Bonnett combines a deep knowledge of history and contemporary geopolitics with a seasoned travel writer's eye for the telling detail, as he gives us a tour of our terrifying but often beautiful new world.' Joshua Keating, author of *Invisible Countries*

'An ambitious journey by wing, sail, rubber and road to find the lost, emerging, off-limits and artificial islands of our fast-changing world. Once again, Bonnett respectfully drags geography back to its roots.' Bradley Garrett, author of *Bunker: Building for the End Times*

'Bonnett's reporting of islands new and ancient – from trash islands to military islands to new environment-trashing "ultrastar" islands to approaching-extinction islands – is a well-researched and open-handed cautionary tale for our times.' Dan Boothby, author of *Island of Dreams: A Personal History of a Remarkable Place*

'Sheer vulnerability and bold architecture live cheek by jowl in this book. If islands did not exist, we would have to invent them. And now we do. *The Age of Islands* helps us understand how and why.' Godfrey Baldacchino, University of Malta; president of the International Small Islands Studies Association

*

By the same author

Off the Map

Beyond the Map

New Views: The World Mapped Like Never Before

What is Geography?

How to Argue

THE AGE OF ISLANDS

IN SEARCH OF NEW AND
DISAPPEARING ISLANDS

ALASTAIR BONNETT

Atlantic Books
London

First published in hardback in Great Britain in 2020 by Atlantic Books,
an imprint of Atlantic Books Ltd.

This paperback edition first published in Great Britain in 2021
by Atlantic Books.

1 2 3 4 5 6 7 8 9

A CIP catalogue record for this book is available from the British Library.

E-book ISBN: 978-1-78649-811-3
Paperback ISBN: 978-1-78649-812-0

Printed in Great Britain

Atlantic Books
An imprint of Atlantic Books Ltd
Ormond House
26–27 Boswell Street
London
WC1N 3JZ

www.atlantic-books.co.uk

CONTENTS

Part Three: Future

Introduction

THIS IS THE age of islands. New islands are being built in numbers and on a scale never seen before. Islands are also disappearing: inundated by rising seas and dissolving into archipelagos. What is happening to islands is one of the great dramas of our time and it is happening everywhere: islands are sprouting or being submerged from the South Pacific to the North Atlantic. It is a strange rhythm, mesmerizing and frightening, natural and unnatural. It is imprinting itself on our hopes and anxieties: the rise and fall of islands is an intimate and felt thing as well as a planetary spectacle. I want to navigate this new territory and try to grasp what it tells us about our relationship – our vexed love affair – with islands.

This is the story of that adventure. It won't be plain sailing. I know that for certain now because I'm writing this in Nuku'alofa, the slow-moving, weather-battered capital of the Kingdom of Tonga, and I'm feeling just as tired as any of the sad-eyed dogs that hunker on the hot and empty road outside. This morning the wind blew unexpectedly hard, and 15 kilometres from shore the hull of the unexpectedly small motor launch on which – many weeks before and many thousands of kilometres away – I'd booked a passage to a newly emerged and as-yet-unnamed volcanic island began to fall in sickeningly slow blows, hammering every valley between every green wave. 'We must turn back,' hollered the captain, the faded tattoos of whales and dolphins writhing with the spray along his bare arms and chest.

So, yet again, I'm holed up, WhatsApping friends and family: 'Didn't make it to my island.' I've come 17,700 kilometres for nothing. A cyclone is hitting this patch of the Pacific tomorrow and I guess I will never reach that impossible fleck on the horizon.

'My island'. What a strange conceit. Islands get under your skin like that: splinters of longing, or escaped territories, they lodge themselves deep. As the gathering storm spits down its first heavy drops, I treat myself to another splash of whisky and trawl my memory, not for the first time, for what set this long and often lonely journey in motion. I remember my seventeen-year-old daughter standing in the kitchen, toast-in-hand, wise, steady and unimpressed. 'You're fundamentally dumb,' she warned me with icy authority, adding: 'All you're doing is globalizing your male menopause.' But then she smiled gloriously: 'I want to come!' Others were less generous, narrowing their eyes in the presence of some unfortunate but undefined species of post-colonial self-indulgence.

Yet chasing these scattered and unmapped points of change feels urgent to me. I keep waking suddenly in the small hours, obsessed with some wayward, unanchored detail, only calmed when I have scribbled out a map or illegible note. I guess I need to cool down and tell this story slowly, to work out why the rise and fall of islands matters.

There is no place better to start than the South China Sea. To the north and west the coastlines of China and Vietnam bulge into its warm waters; to the south and east lie Malaysia and the Philippines. This is one of the world's great trade routes – said to be worth $5.3 trillion a year – and it is one of the cockpits of contemporary geopolitics. The Spratly Islands, the once-pristine and untouched reefs and tiny islands that sprinkle this sea, have been horribly mutilated: squared off and concreted over; a dozen or so have been crammed with military firepower and turned into audacious forward placements in a new cold war. China is bolting together the majority of these Frankenstein islands and it is winning control of the entire sea.

Satellite and aerial images show how the reefs are latched on to by long black snake-like pipes that curve through the water. They wend back to boats that are grinding up the sea floor – sand, coral, crustaceans, everything – into building material. This marine paste is squirted onto the island. Later come the concrete mixers, the airstrips, naval harbours and the missile silos. One of the latest victims is Johnson South

Reef. It has been snared by an inseminating predator. In its early stages it is bulked up. Later it will be squared off – a hostile alien in a beautiful blue sea.

The tragedy of the Spratlys has been spread across headlines in East Asia for some years. In the coming decades much bigger and more peaceful Chinese islands will grab the world's attention. Spectacular new leisure and entertainment islands are emerging just minutes from the shore of a number of coastal cities. Like the artfully shaped new islands sculptured in the Gulf States, these are sites of turbo-charged consumerism. However, since they are made by gouging out the seabed and planting rows of offshore and improbably shaped, air-conditioned hotels, these apparently carefree shopping and holiday destinations can be just as environmentally damaging as their military cousins.

Our power to reshape the planet is stark on new islands. Each of them shouts: 'Look what we can do!' But the age of islands has another face. New islands are rising up as old ones are going under. Today the spectre of disappearance stalks low-lying nations. Thousands of the world's islands are only centimetres higher than the surrounding sea and most are shrinking year by year, month by month. The roll-call of the vanished is already a long one. The rate at which dredgers and engineers can fabricate new islands is increasing but so is the speed at which natural islands are being swallowed up.

En route to the capital of the Solomon Islands for yet another conference on climate change, the then UN Secretary General Ban Ki-moon peered out of his aeroplane window and saw what may, at first glance, have appeared to be a couple of undersea reefs and, in the background, some small islands. In fact, it was the remnants of one large island that had been almost completely swallowed up, with only the highest ridges left. About a dozen islands in this part of the Solomons have gone the same way. Islands today feel fugitive and uncertain: an atmosphere of doubt surrounds them. Their stories hold a mirror up to our alarming era.

Islands are changing fast but they have a primal allure. I love islands. They offer the possibility of newness, of hope. Staring at the white, lifeless heap that is Johnson South Reef or the vestiges of the Solomons, that might sound very odd. But the idea of utopia clings to even the bleakest island. The first image of 'utopia' was an island. It is telling how insistent Thomas More was in the book usually simply titled *Utopia* – his travel fantasy that gave us the word – that Utopia had to be an island. More tells us that the founder of this uniquely perfect realm, King Utopus, 'made it into an island'. Originally it was part of the mainland but Utopus 'ordered a deep channel to be dug' in order to 'bring the sea quite round'. Only in this way could a flawless and completely new place be born. Utopia is a space apart, a jewel in the sea, a distant sight towards which one longs to steer.

More describes the island as 'not unlike a crescent: between its horns, the sea comes in eleven miles broad, and spreads itself into a great bay, which is environed with land to the compass of about 500 miles'. It's easy to imagine sailing into that generous harbour. One of the alluring things about islands is that we can picture them whole in our mind's eye. Hence we can imagine them perfect – complete and completed.

Anyone who travels to new islands has to deal with hope. Not timid, doe-eyed hope but outrageous, gleeful, turbulent, confident hope. It's there in the fast-mutating island polders of the Netherlands and the off-kilter leisure islands of the Gulf States and China. Despite the fact that most new islands are environmental disasters, it's still, perversely, impossible to detach them from hope. So it seems inevitable that I devote the last section of this book to *future* islands – places that are likely to be unveiled over the next few decades or so.

Another memory is rising to the surface. My first 'new' island. I went there a couple of years ago. I recall it like the face of an old friend. I need that memory now for the rain and wind are pounding the roof. Best not to listen as the palm trees clatter and twist, their limbs snapping and skittering skywards. Many people on Tonga are spending tonight in flimsy canvas tents donated by aid agencies in the aftermath of the last cyclone, often camped in their own wet gardens. Casting back for comforting memories, I retreat to a happier place.

A light heave on the oars. The water is calm, silky. This is how all island journeys should be. With that last pull, the pretty green rowing boat I've rented for the day grates a few inches onto an underwater rim of hefty round stones. I splash out and begin busily surveying, pushing a yellow steel tape measure through a tangled knot of shrunken alders and over the grizzled corpse of a putrefying sheep. (How did it get here?) My unnamed island is 19.5 metres long and 10 metres wide. High above it heavy black power cables sag across the width of the loch: dark arcs drawn on a summer sky. It is a windless day.

The island is one of many on Loch Awe, a freshwater lake 40 kilometres long in the west of Scotland. I had no inkling that it marked the start of something. My yellow tape measure, wielded with faux-professional aplomb, is nothing more than a protective talisman, fending off the pointlessness of a wayward mini-break. Squatting down on this islet's western shore beside a mini-vortex of plastic rubbish – coagulated food packaging and fishing lines – I baffle myself with questions: 'Why have I come here? Why do islands lure me?' Staring down at all that plastic, other questions soon come: 'What is happening to islands? Why, today, are we building so many of them and misshaping many more?'

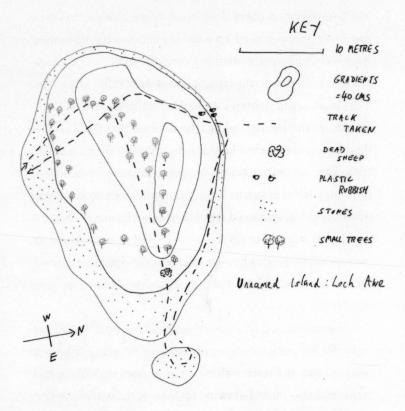

KEY

|——————————| 10 METRES

GRADIENTS
= 40 CMS

- - - TRACK
TAKEN

DEAD
SHEEP

PLASTIC
RUBBISH

STONES

SMALL TREES

Unnamed Island: Loch Awe

This cloud of question marks might seem out of place in such an anonymous and peaceful place. But my island holds a secret: it was built by people. So were nearly all the other islands I can see from its shore. The two dozen or so that are still visible in Loch Awe were all constructed somewhere from 2,600 to 600 years ago. Surrounded by high hills, back then people used the rivers and lochs as their transport routes. They were water-based in almost everything they did and island-building allowed them to live on their economic and political 'main street'. Logs were driven into the shallows and large stones placed on top. Communal round houses were then built on the islands alongside small pens for pigs and goats. These ancient artificial islands are called crannogs. Only handfuls have been excavated. Scotland has about 350 existing examples; Ireland has many times that number and there are similar ancient lake islands in dozens of other countries. They are intriguing places: perplexing yet immediately understandable. Humans have an insatiable curiosity about islands and a deep-rooted desire to shape and create them.

After that trip out on Loch Awe I drove back home to Newcastle, the city in England's far north where I have lived for thirty years, and tried to put my feelings into some kind of order.

I began doodling all sorts of island shapes, just like I did as a boy: fat and fiddly ones; ones with lovely sinuous inlets; ones with villages and ones with mountains; ones with caves and treasure. I also started drawing up a list of the world's newest and most rapidly changing islands – both the natural and unnatural. These came from conversations at work, back in Newcastle University's Geography Department (where I first got word that the main part of Svalbard, in the high Arctic, was revealing itself as two islands as the ice sheets melt away), and news items (latest pictures on the bulked-out and militarized Spratlys in the South China Sea and a menacing-looking new volcano rising from the sea north of Tonga). I also relied on readers of my previous books on 'off the map' places. Had

I heard of the new artificial islands in Korea? Did I know about the 'Trash Isles', or islands that are poisoned, exploding, becoming uninhabitable, crowded with giant crabs?

So many islands. I'm not sure it helped to find a copy of the *Island Studies Reader* and learn that there are 680 billion islands on our planet. It turns out that figure includes 8,800,000 islets and 672,000,000 rocks. I wonder who counted them. It sounds like guesswork. I was becoming overwhelmed: unsettled by islands' fractal endlessness and the fast pace of change.

Trying to clear my head, I stopped spending my evenings on Google Earth (which is often years out of date when it comes to new and disappearing islands) or checking my email (*ping*: 'I saw this on the BBC News App and thought you should see it: "The island that switches countries every six months"'). I wanted ideas that would anchor me. I kept coming back to an idea that proved seaworthy and still guides me, namely the Janus-faced nature of modern islands: they are both frightening and beguiling; they offer security but also vulnerability. Here are some of my more cogent scribblings from when I was just back from Loch Awe.

Islands = crisis: the drama of so many issues – climate change, species loss and extinction, overpopulation, nationalism and pollution – is played out with a special intensity on islands. The disappearance of an island occasions genuine grief, a real sense of loss, in a way that ordinary flooding does not. When an island goes it is like a complete thing, a whole nation, has been eradicated. Islands are often small places but they pack a big punch. Conquering or creating new ones is a big deal. Countries are greedy for them – in part because they can claim 200 nautical miles of territory from the shore of every single one. They offer a radical leaping outwards of national power. If you are looking for places that are intensely occupied by military firepower or have been bombed to nothing, then islands are the places to go.

Islands = freedom and fear. They seem tailor-made for experiments and a fresh start. Perhaps that's part of the tingle when the boat nudges the shore, the possibility that this is a new world where things, finally, can be made right. The twenty-first century is throwing money and ideas at islands. The rich like them because they offer security and status. But in an era of rapidly accelerating sea-level rise and worsening storms, islands are fragile. They are the first places to be abandoned. The dream becomes a nightmare and the island a prison. Islands are often used as dumping grounds for the unwanted. They lure us but they can easily and quickly become places of dread.

The windows have started to shudder and the wall plugs just sparked. 'No way you're flying out this week,' my Tongan captain forewarned me. As the cyclone gathers, I'm shrinking and hunching, trying to get small. My notes about islands suddenly feel like very thin gruel. They may contain some truth but they don't feel vulnerable enough. Our relationship to islands goes well beyond political and ecological headlines or clever paradoxes. I try another memory; I reach back, much further.

I'm standing with my brother and sister in an old wood – it is called Wintry Wood – at the northern end of Epping, the town on the eastern outskirts of London, where I was born and grew up. I've got my bright red wellington boots on; sometime later one of them will be sucked down and lost in a nearby patch of bog. Before us – Paul and Helen and me, the youngest – there is a dark, quiet pond hazy with flies and in that pond is an island. The pond and the island must be very old but they do not look natural: dug out for reasons long forgotten and of no interest to us. The island has our full attention: it is our destination. It is maybe 100 square metres in size and dense with beech and silver birch trees. Thin, finger-like branches reach down and paddle the water, beckoning children. Generations have taken up the invitation. We edge sideways down one particular muddy bank where an uneven causeway of sticks has been strewn – branches and twigs thrown into the water by those who came before us. It's likely at least one of us will get a welly full of stinking, leaf-matted water and have to beat a wet retreat. But not this time, not in this memory. I can't help smiling: we've all made it, holding hands across the tricky bits.

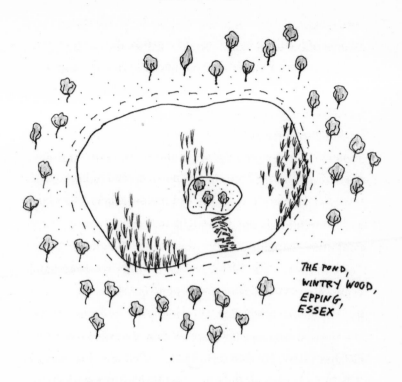

THE POND,
WINTRY WOOD,
EPPING,
ESSEX

But once on the island, what to do? There's a glitch in my happy memory. The triumphant three stand aimless, patrol proudly, going nowhere. The island offers elation and restlessness; soon we're splashing back to the mainland with a sure sense of accomplishment and a tale to tell.

The voyage to an island stays with you, moored inside for reasons that are hard to explain. I think I knew I wanted to become a geographer when I came across a book called *Topophilia* by Professor Yi-Fu Tuan that recognized the depth of this mystery. Tuan wondered why 'Certain natural environments have figured prominently in humanity's dreams of the ideal world.' And he names them: 'the forest, the seashore, the valley and the island'. His list pointed to the natural advantages of these landscapes and so refer-enced the work of another thinker interested in why we are attracted to some landscapes and not others. This was Jay Appleton, a poet-geographer who died in 2015 and who worried away at the question of why people find high moun-tains and islands alluring. We stumble around trying to pin down our feelings about landscapes, said Appleton, using words like 'happiness' and 'grief', even though we know they don't really fit. We are, he said, 'using a second-hand terminology to describe a relationship which we do not properly understand'. For Appleton it's a relationship that goes beyond language for it is based on atavistic instincts of fear and safety. He devised what he called 'prospect-refuge theory' to impose some order on this uncertain terrain. Humans, he argued, have an 'inborn desire for places where they can assess threats from a place of safety'.

It makes some sense and our fascination with islands, especially small ones that we can observe all at once, is, in part, explained by Appleton's 'prospect-refuge theory'. A less scientific take on this allure is found in D. H. Lawrence's story 'The Man Who Loved Islands'. Based on the real-life island-hopping of the novelist Compton Mackenzie – who, in the 1920s, leased or bought a series of ever-smaller British islands (first Herm then Jetou, which are Channel Islands, then the Shiant Isles in the Outer Hebrides), each of which was his sole domain – Lawrence wrote that 'an island is a nest which holds one egg, and one only. This egg is the islander himself.'

During a long, sleepless night and all morning Cyclone Keni bruised Tonga. Like everyone else, I peep out of my door in the afternoon when it feels safe. The windows of almost every building had been boarded up so there is not much glass on the streets. Taking to my hire car, I soon come across schools and government buildings marooned by new lakes, with kids whooping and splashing in the warm water. Although large white defensive rocks are piled on some northern shorelines, this water has poured straight down from the sky and now it lies all around in shining sheets. It is so humid that I have to keep on wiping tiny rivulets running down my glasses. After thirty minutes driving I end up at the island's most western point. An information board proclaims it was here that Dutchman Abel Tasman 'discovered' Tonga in 1643. This afternoon, the spot is colonized by a giggling and friendly group of gay and transgender youths. They are dancing round a tiny speaker, enjoying the return of some sunshine. Their first question is the same question all Tongans ask of strangers, asked in the same quiet way with an uncertain, shy smile: 'What do you think of Tonga?' I can tell they are steeled for a put-down, expecting to be hurt. Each time I'm asked this kind of question I freeze. I can't tell the truth because they wouldn't believe me. The truth is that l think it is beautiful, unique, and it feels like a huge privilege to be here. We all laugh, I mutter something about the weather, and I guess they go away believing that my opinions about Tonga can be reduced to the words 'it is very wet'.

Weeks later, years later, I wish I could go back and explain, get it right. The least I can do is to dedicate this book to them and to all those who love islands.

PART ONE

RISING

Why We
Build Islands

N A DARK bar on the shores of Loch Awe a tall, beery fellow leaned into me and slowly explained that the crannogs – the ancient homesteads sprinkled in the lochs of Ireland and Scotland – were the very first artificial islands. I nodded meekly. It seemed likely and he was staring at me with red-eyed certainty. If I'm ever up that way again I may have the courage to lean back and put him right. The truth is that artificial islands are found across the world and that trying to claim any one as 'the first' is like trying to locate the first firepit or the first hut. Although often overlooked today, they are just too common to be easily or usefully tracked down to a single original source.

What are they for? Sifting through the layers of island-building history, the main reasons why people built them can be organized as follows: for defence and attack; to create new land for homes and crops; as places of exclusion; as sacred sites; and finally a rag-bag category of islands for lighthouses, sea defence and tourism. If we drill down into each of these purposes, we start to see continuities to our modern age of islands but also differences, not just in terms of number and size but in how they are used. For the majority of the world's new islands have no pre-modern predecessors. These are the rigs and turbines, dedicated to oil, gas and wind power extraction, that dot so many horizons.

Defence and attack

Many of the reefs of the South China Sea have been bulked out and squared off to house missile silos, naval docks and runways. Although there is a long history of new islands born of strife, the oldest have nothing to do with sabre-rattling. In the Solomon Islands, the Lau fishing people built about eighty islands in a sheltered lagoon by paddling out – year after year, for centuries – and dropping lumps of coral into the water. The Lau built these islands to escape attack from mainland farmers. Many are still inhabited. Their defensive function has ceased to matter but they still offer protection from wild animals and malarial mosquitoes. Elements of this story can also be heard on Lake Titicaca in South America where another fishing community, the Uros, built a similar number of islands many miles from the shore in order to be safe from aggressive neighbours. Unlike the Lau's solid structures, the islands of the Uros are made of reeds and float. This design reflects the building material to hand but also allowed the islands to be moved if under threat. Reed islands last about thirty years and need to be continuously remade. The Uros maintained these woven structures across hundreds of years. Today they are much closer to the shore and attract tourists from all over the world.

Ancient defensive artificial islands were small, occupied by families not soldiers, and never had much, if any, weaponry. In Europe the construction of more robust and professional artificial island fortresses began in earnest from the seventeenth century, and over the next three hundred years imposing stone forts were built on numerous reefs and sandbanks, usually to guard important ports. Some of the grandest were built by Louis XIV, such as the horseshoe-shaped Fort Louvois. Foundations for Fort Louvois were sunk into a muddy rise in the sea near Rochefort on 19 June 1691. At high tide it still looks startling: a castle rising from the water. In fact, Louvois saw only brief bouts of active military service. The last came on 10 September 1944, when it was shelled and briefly occupied by the fleeing German army.

Like a lot of militarized islets, the history of Fort Louvois is largely one of inactivity. Their main role has been as deterrents: they look big and bold in order to make invaders think twice. Peter the Great, having founded St Petersburg, sought to defend his creation with a series of spectacular sea forts. The first was Fort Kronshlot, built in shallow water during the winter of 1703. The most famous of the Petersburg forts is Fort Alexander, an immense oval begun in 1838. Fort

Alexander was big enough to accommodate 1000 soldiers and 103 cannon ports. Like so many other dramatic offshore forts, Fort Alexander quickly became outmoded and, in military terms, useless. Having been demoted to a storage depot, it was given a new lease of life in 1897 when it became home to the research laboratory of the Russian Commission on the Prevention of Plague Disease. For twenty years this isolated, stone citadel caged a variety of animals used in plague experimentation, including sixteen horses whose blood was used to produce plague serum.

Their military life may be brief but the story of any well-built sea fort is rarely a short one. Cut off from the bouts of demolition that afflict the mainland, they often last a long time and see a range of uses. Many European nations are still trying to work out what to do with the military islands that freckle their coasts. Many stem from the busiest period of European island-building, which came in response to the threat of Napoleon and his heirs. Napoleon built his own islands, the most striking of which is Fort Boyard, an austere oval – resembling a giant napkin ring cast into the sea – that was built between 1809 and 1857. For many years it lay empty, but in the 1990s Fort Boyard began a new life as the setting for a French 'escape the castle' television game show that has had a number of international spin-offs.

Across the English Channel, Victorian sea forts are similarly intriguing but have been equally hard to find a modern use for. Occasionally they come up for sale, such as 'Number

1, the Thames', an address also known as Grain Tower Battery. This bizarre hodgepodge of Second World War gun emplacements stuck onto a mid-Victorian military island sits in one of the widest reaches of the Thames estuary. In 2014 it came up for sale for £500,000. This sounds like a meagre asking price given that islands are usually one of the most expensive types of property. However, while old sea forts may look good, they come with colossal maintenance bills. A similar problem kept down the price for the five forts built to defend Portsmouth in the 1860s. With 4.5-metre granite walls and armour plating, even when newly built they combined magnificence with obsolescence: at the very moment of their completion, the threat of the French invasion they were built to repel disappeared. In 2009 three of the forts were acquired for conversion, one into a museum (Horse Sand Fort) and two into luxury hotels (No Man's Fort and Spitbank Fort). After huge investment, ten years later all three were being advertised for sale again.

Some fort-islands are so large and remote that their commercial opportunities are limited. Fort Jefferson is 19 hectares in size and 109 kilometres west of Key West in Florida. The largest brick building in the Americas, it was constructed in 1847. After being used to blockade the Confederate States during the Civil War, it had a limited life as a military prison and was abandoned in 1906. It is now an out-of-the-way tourist attraction set within the Dry Tortugas National Park, one of the most inaccessible of America's National Parks.

Many sea-forts do not even function as quirky tourist destinations and lie totally abandoned, breaking slowly apart under a weight of weeds. The hexagonal Fort Carroll is one such place. It lies in the Patapsco River in Maryland. Built to defend Baltimore in the 1840s it was bought by a family in 1958 and left to become what it is today, a stone wilderness and home to thousands of nesting sea birds.

Nineteenth-century fort-islands were built of stone or brick. During the First and Second World Wars military engineers began to use metal and the result was an array of raised – and, not long after, rusting – structures. The most famous is Sealand in the English Channel. On 2 September 1967, retired army major 'Paddy' Roy Bates climbed onto it and declared it was an independent country – a claim that is maintained by his descendants to this day. Sealand was one of a range of sea forts built off the English coast in 1942–43 that resemble oil and gas rigs. Some, like Sealand, have two rotund supporting legs while others rise on thin stilts and have a number of interconnected platforms. Assailed by stormy, salty seas, examples of the latter have not fared well. A few dilapidated examples linger on and had interesting afterlives, such as Shivering Sands Army Fort, which in 1964 was turned into a pirate radio station by the eccentric politician Screaming Lord Sutch.

After the Second World War, island-building shifted to the Pacific. The USA began reshaping atolls for military purposes. Johnston Island was bulked from 18 to 241 hectares in order to accommodate a landing strip. Today it is a long, unnatural-looking rectangle. At its peak about 1000 personnel were stationed there. The island was used for nuclear weapons testing in the 1960s and has a 10-hectare landfill full of toxic material, including drums of Agent Orange from the Vietnam War. To add to the poisonous brew, the island hosted an incineration plant for chemical weapons, including Sarin nerve gas.

New land for homes and farms

The most common form of ancient artificial island is a small homestead. The crannogs are one example. The homes of the Ma'dan, or Marsh Arabs, are another. Since the fourth century BC the Marsh Arabs have been building floating islands made of reeds in a junction of the Tigris and Euphrates rivers in Iraq. Their way of life was almost destroyed when Iraq's president, Saddam Hussein, drained the marshes. Since Hussein's fall, there have been determined attempts to restore the marshes, and small groups of Marsh Arabs have chosen to return, fitfully reconstructing their former way of life.

Numerous homestead islands may accommodate a lot of people but each is a small and simple affair. A pre-modern counter-example is the island city of Tenochtitlan, the site of modern-day Mexico City. The invading Spaniards couldn't believe their eyes when they saw it. They called Tenochtitlan 'a very great city built in the water like Venice'. Bernal Díaz del Castillo, writing in 1576, told how the Spanish marvelled at Tenochtitlan's size and beauty and at how it was a:

wonderful panorama, as picturesque as it was novel [...] on account of the great towers and temples and buildings rising from the water, and all built of masonry. And some of our soldiers even asked whether the things that we saw were not all a dream.

Tenochtitlan was built by the Aztecs in the early fourteenth century and was home to around 500,000 people. It was semi-artificial, being extended from a natural island to spill out over several islets establishing a 13-square-kilometre platform connected by 20 kilometres of canals and raised roads. Historian Gerardo Gutiérrez tells us that 'Moving through the city of Tenochtitlan would have involved a combination of canoes and walking through a complex network of streets and alleys connected by hundreds of bridges.' Tenochtitlan linked together numerous artificial farming islands called chinampas. Chinampas are sometimes called 'floating gardens' though they don't actually float; they are made by staking a reed fence to the lake bottom then piling on material until an island emerges. In Tenochtitlan, with the help of the chinampas, farming was urbanized, with hundreds of rectangular field-islands, all artificially created, linked in rows and bedded into the fabric of the city.

The 'floating city' is one of the many labels given to Venice, whose 118 islands are knitted round a network of canals and bridges. From the fifth century AD settlers fleeing raiders from the north began living on the region's marshes.

Later generations drove wooden stakes into the mud and created basic wooden platforms and buildings. From these simple beginnings the city of Venice emerged, perfecting the art of the semi-artificial island. To give an idea of the effort involved it is enough to note that, in 1631, to build Venice's church of Santa Maria Della Salute, 1,106,657 4-metre wooden stakes had to be piled into the water.

The name of Venice is conjured again and again, like a charm or talisman, in modern residential island developments. The Venetian planning model of multiple canals and houses with direct access onto the water has been rolled out worldwide. Coastal reclamation for these schemes helps explain why – despite sea-level rise – since 1985 the world has gained more land from the sea than it has lost: an area, according to the Dutch research group Deltares, about the size of Jamaica. 'The Venice of America' is Fort Lauderdale, 40 kilometres north of Miami. Once a country town, from the 1910s Fort Lauderdale began to be transformed by entrepreneurs who realized that maximizing waterfrontage for newly built homes would create upmarket sales. Canals were built, land reclaimed and soon residential communities such as Las Olas Isles and Seven Isles were attracting buyers willing to pay over the odds for an exclusive address that offered privacy, security, views over water and quick access to their own boat. Although these new communities were usually advertised as islands, they were nearly always 'finger islands' – long, thin peninsulas joined by circuitous roads to the mainland. Finger-island development was to spread up and down the coast of Florida and, later, to the more prosperous parts of the coastal world.

Another influential set of Florida islands lies between Miami and Miami Beach in Biscayne Bay: the six 'Venetian Islands' built in the 1920s and 1930s and connected by a highway. Their names pay homage to their Italian forebears, such as San Marco Island, San Marino Island, and Di Lido Island. They showed that finger islands were not the only way to make money, and that real islands (albeit connected by a highway) could also offer canny developers premium returns.

The 'islandization' of coastlines is, for the most part, a form of suburbanization. The new island suburbs are scattered wherever water, land and money collide. As they grow they join up, creating an uninterrupted string of built-up water-facing townships that stretch along the seaboard. In some places, like Australia's Gold Coast, this has resulted in the coastal landscape being transformed from a natural beachscape into a long chain of artificial residential island developments. There is a pecking order, of course: the standalone, 'real' islands usually have higher property values than the finger islands. The Gold Coast's Sovereign Islands are its most expensive address. Reclaimed from sandbanks and mangroves, this gated community comprises six connected islands with a single bridge to the mainland. It was formed by dredging 2.3 million cubic metres of sand from adjacent waterways, a process that both built up the land and cleared a channel deep enough for the largest of luxury boats. Many of the houses on the Sovereign Islands are like palaces. They are well away from prying eyes but they are ostentatious. With names like Palazzo di Venezia or Château de Rêves (which became famous for having its swimming pools lined with 24-carat gold tiles), they are confident yet safely distant – loud statements of wealth that do not wish to be disturbed. Today the best-known examples of this kind of development are in the Gulf States. I'll be exploring some of Dubai's most outlandish examples later, when I take a trip to the most bizarre of them all, The World.

Some planning experts claim that we are transitioning to a feverish, hyperactive state in the creation of ersatz island locations. Two geographers from the University of Bristol, Mark Jackson and Veronica della Dora, argue that the 'worldwide phenomenon of the artificial island has become a key defining imaginary and material form of 21st-century development visions'. Jackson and della Dora suggest that 'urbanizing coastlines' are seeking to 'ornamentalize' themselves. But living cut-off from the mainland is not just about decoration. It's a form of self-exclusion. Islands are safe havens: away not just from noise, crime and crowds but from dirt and disease. Many small islands managed to stay Covid-free during the pandemic. The virus has become yet another reason why people might want to retreat into defensible territory, tucked away behind a barrier of salt water.

HOW THE MALDIVES IS BUILDING A FUTURE

All of the 1190 islands that make up the Maldives are projected to be underwater by the year 2100. The Maldives are fighting back with a flotilla of new islands, including an island city, floating islands for tourists and an island that landfills the endless garbage created by its tourist-based economy. The country's tourist slogan 'Maldives – Always Natural' could scarcely be more misleading.

The centrepiece of the fightback is Hulhumalé, branded the 'City of Youth' and the 'City of Hope'. It has been built 2 metres above sea level on a coral reef. Phase one is now almost complete and, in 2013, had a population of 30,000. The target is 60,000. It's a 188-hectare urban rectangle with rows of anonymous apartment blocks. The 240 hectares of Phase two offers taller, more glitzy buildings and is designed to house a further 100,000. The idea is that Hulhumalé can become a safe home for about half the population of the Maldives. Where the rest will go is not so clear, though there is no shortage of plans, including mass relocation to India.

In the meantime the government is capitalizing on the Maldives' 'natural' beauty while it can and, in 2016, passed a law allowing foreigners to buy land in the country, as long as they had $1 billion to invest and 70 per cent of the land they bought was reclaimed from the sea. Rich foreigners can also buy property in 'The Ocean Flower', 185 floating villas built by Dutch Docklands and arranged in the shape of a Maldivian flower. Other Dutch Docklands plans given the green light include Greenstar, a star-shaped floating hotel and a floating eighteen-hole golf course that the promotional bumf claims 'will feature many water hazards and 360-degree ocean views'.

Water hazards and ocean views are not in short supply in the Maldives, though the government works hard to not shadow the island's carefree tourist image with the sombre and unpicturesque reality of living with sea-level rise. Well away from the tourists' floating playgrounds a very different sort of artificial island has been built to hide the island's trash and its most polluting industries, such as cement packing and gas bottling. The island is called Thilafushi. It was once a blue lagoon but it is now a squared-off landfill, where 300 tonnes of rubbish are shipped every day.

Islands of exclusion

Islands have been built to deposit all manner of undesirables. Venice has an archipelago of dumping grounds: San Lazzaro degli Armeni started life as a leper colony; San Giorgio in Alga was for political prisoners; and San Servolo housed the city's insane asylum. There was also a number of plague islands, such as Lazzaretto Vecchio and Poveglia where it is said so many died that half the soil is made up of human remains. Creepy legends continue to stalk Poveglia. The story goes that, when a mental hospital was opened on Poveglia in 1922, one of the doctors went mad and began torturing and killing patients before they rebelled and he was thrown off the bell tower. A more recent tale has it that in 2016 five American tourists tried to spend the night on Poveglia. It didn't go well. As soon as darkness fell a terrible presence began to harry them. Their screams were eventually heard by a rescue crew of passing firefighters.

Some islands of exclusion were also windows to the outside world. Dejima was built in Nagasaki bay in Japan in 1634 to contain foreigners but also to provide a safe place to meet Portuguese and then Dutch traders. The fan-shaped, 9000-square-metre island was ordered to be constructed by the local shogun because he feared the presence of Europeans on Japanese soil would encourage the spread of Christianity. Life on the island was strictly monitored. During its long period of Dutch occupation, the twenty or so foreign residents were watched over by a staff of more than fifty gatekeepers, nightwatchmen and other officials. The last Dutch 'overman' of the island left his post in 1860, just before the emperor began opening Japan to the outside world. Although now integrated into the rest of Nagasaki, a reconstruction project has been restoring some of Dejima's buildings, and in 2017 the newly recreated old bridge to the island opened with members of the Japanese and Dutch royal families in attendance.

Ellis Island, which sits in Upper New York Bay, is a semi-artificial island that grew from 1.3 hectares to 11 hectares, mostly with material derived from ship ballast and debris from the building of the New York City subway system. It served as a site for a gibbet, a fort, an ammunition depot and finally, in 1892, an immigration processing facility that processed more than 12 million applicants over the next sixty-two years. Since first- and second-class travellers were allowed to go straight to the mainland, most of the people it dealt with were third-class passengers. From Ellis Island, anyone whose application for entry to the USA was refused (with disease being one of the main reasons for refusing entry) would not have the option of evading the immigration authorities and could more easily be sent back home.

Sacred islands

An island is a space apart. One of the ancient reasons they were built was to create sacred sites reserved for rituals and high priests. This doesn't happen today but there remains something otherworldly and extraordinary about all islands, artificial or natural. One of the most remote and intriguing examples is Nan Madol, a complex of about a hundred artificial islands reserved for priestly and elite ritual found off the island of Pohnpei in the Federated States of Micronesia. Inevitably dubbed the 'Venice of the Pacific', Nan Madol was built from the eighth century and many of its coral stone walls and buildings have survived, lacing the shallows with an overgrown and tumbling geometry of ruins. Said to be built by twin sorcerers, Nan Madol had a complex function, providing homes for an aristocratic social caste and a cluster of fifty-eight islands dedicated to rituals surrounding the entombment of the dead.

The act – the effort – of creating a sacred island can itself be a form of veneration. This appears to have motivated the building of a number of religious islands such as Our Lady of the Rocks in Montenegro. This island, which hosts a small lighthouse as well as a church, was built by generations of sailors sinking ships loaded with stone around a rock where an image of the Madonna and Child is said to have appeared on 22 July 1452. A festival is still observed on this date: a convoy of small boats connected and decorated with branches and banners rows across and then drops stones around the base of the island.

Sacred islands are places of legend. Sometimes, like Chemmis – the fabulous floating island that held a temple to Apollo described by the ancient Greek historian Herodotus – they belong only to the world of myth. But the relationship between myth and reality has often been confused, especially when legendary sacred islands are replicated as an act of homage. This brings us to the three most important sacred islands of Chinese mythology: Penglai, Fangzhang and Yingzhou. They were searched for by Chinese emperors because, as well as being retreats of the gods, they were said to hold the elixir of immortality. Expeditions to find them returned with reports of sightings. According to the second-century BC *Records of the Grand Historian*, 'All the plants and birds and animals of the islands were white, and the palaces and gates were made of gold and silver.' Such was the utter conviction that they were real that one emperor sent a colony of young men and women to take possession of them.

Penglai, Fangzhang and Yingzhou were not only searched for, they were reproduced. Imitations of the islands were built within lakes in the palace gardens of many emperors, including in Beijing's twelfth-century imperial garden along-side the Forbidden City. These mirror islands became part of a well-known landscape-design pattern known as 'one lake, three mountains', in which the 'mountains' were the sacred islands, each topped with a dramatic temple. The Fairy Isle of Penglai is the best known of the three islands, and visitors to the Chinese city of Penglai can today enjoy a heritage park containing one lake and three 'mountains'. Although some have described the result as tacky, the historical journey from legend to ancient reconstruction to modern theme park is a powerful lesson in how sacred islands retain their hold on the imagination. Frequent mass hallucinations seen off the coast of Penglai drive the point home. One news item, relayed by China Broadcasting in 2006, ran:

Thousands of tourists and local residents witnessed a mirage of high clarity lasting for four hours off the shore of Penglai ... Mists rising on the shore created an image of a city, with modern high-rise buildings, broad city streets and bustling cars as well as crowds of people all clearly visible ... Experts said that many mirages have been recorded in Penglai, on the tip of Shandong Peninsula, throughout history, which made it known as a dwelling place of the gods.

Lighthouses, sea defence and tourism

Ancient examples of artificial islands used for lighthouses and sea defence do exist but they are rare. An artificial lighthouse-island was built in the first century AD at Portus, Rome's harbour. It is said its foundation is the concrete-filled ship that in AD 37 carried the Egyptian obelisk that now stands in St Peter's Square.

From the eighteenth century, lighthouse-building expanded in scale and ambition, creating many semi-artificial islands. By the end of the nineteenth century, completely artificial islands began to dot the shallows and reefs of major ports in seas and lakes all over the world. There are thousands of such islands and in a pinch they are still useful to mariners, though modern GPS-navigation technology has rendered most of them redundant. Lighthouse-building has become a rare event. The last one of any size built in the UK was in 1971, the Royal Sovereign lighthouse, off the south coast, which sits on a sea platform supported by a single pillar.

Protection from flooding is a very old art. One widely used technique was to build high mounds in flood-prone areas. In north-west Europe, these occasional islands were called 'terps' and their remnant small hills can still be seen across the Netherlands, Denmark and northern Germany. Using artificial islands to create barriers between the mainland and the sea is a largely modern phenomenon and, even today, it is rarely the sole function of such islands. Sea defence is *one* of the roles of the polders of the Netherlands but they also serve to provide more land for people and farms. Similarly, Toronto's Port Lands Flood Protection project, which is aiming for a 2023 completion date, will also create a new island, named Villiers Island, for homes, green space and businesses.

The importance of natural barrier islands is often only understood once they are washed away. Restoring barrier islands is a worldwide coastal priority. One of the biggest schemes has been in Louisiana, which has lost 4920 square kilometres to the sea since the 1930s. Today the rapidly eroding Bayou Lafourche barrier island complex is being bulked out and reshaped by a coalition of private and public organizations, including one with the telltale name of 'Restore or Retreat'.

Artificial islands built for mass tourism are a new thing. Older examples of pleasure islands are not uncommon but they were reserved for the very rich and are usually tiny and eccentric. The Palladian Villa Barbarigo at Valsanzibio, south of Padua, features a 'rabbit island' designed to be an attractive talking point as well as a way of keeping rabbits – considered something of a delicacy – safe from foxes. In aristocratic estates, tiny islands were also used as places for tombs and memorials, such as the monument for the eighteenth-century wit and playwright William Congreve that sits in a lake at Stowe, an English stately home and garden. The most marvellous of European aristocratic islands was built by the Duke of Anhalt-Dessau in Germany and is called Stein Island. It featured a rocky island with caves and grottoes, and an artificial volcano that, thanks to subterranean fireplaces, produced real smoke and steam. Completed between 1788 and 1794, it was designed to recreate in miniature the landscape of Naples, complete with Vesuvius. After falling into disrepair, it was restored and reopened in 2005.

In the twenty-first century pleasure islands often combine conference, hotel, marina, theme-park and residential attractions. They are now more like spectacular neighbourhoods dedicated to leisure and shopping than the small, single-attraction islands of the past. They are strange, paradoxical destinations. A taste of their sheer oddness is provided by Indonesia's 'Funtasy Island: the largest Eco Park in the world'. Just 16 kilometres south of Singapore, it caters for Singaporean tourists who can enjoy a mixture of theme parks combined with 328 hectares of 'pristine tropical islands' and an 'unspoiled natural environment'. It is, above all, fun: 'there will be many specially designed spaces that offer worlds of fun dawn to dusk'. Funtasy Island is, apparently, both highly developed and unspoilt – both untouched and luxurious. Its builders have not left nature alone but rather added nature in. They have planted mangroves, created underwater 'structures to attract small fishes and dolphins', and claim to 'plant one coral per visitor'. The importance of the 'eco' in Funtasy Island is indicative of a new trend in island-building. People want nature as a 'sight' or 'experience' and they are more likely to travel to a destination that makes them feel good about the environment. The irony – that a once-pristine reef has been destroyed and built over in order to attract visitors to enjoy 'unspoilt' nature – is unmissable.

Artificial islands are spearheading everything that is bizarre about the twenty-first century. Maybe the fact that they are such 'fun' helps explain how we square the circle. Yet islands built for fun do not have to be relentlessly consumerist or huge. In Copenhagen harbour floats a 25-square-metre wooden platform with a slender linden tree growing in the middle. Free to use and open to all for barbecuing, stargazing and bathing, it is the first of a planned archipelago of similar floating wooden islands supporting various functions: a floating sauna island, floating gardens, floating mussel farms and a floating diving platform are in the offing. After Funtasy Island it feels like a cooling balm, a dose of sanity. Perhaps pleasure islands can also be simple places and rest lightly on the water.

Rigs and turbines

There are plenty of modern artificial islands with ancient ancestors. But there are more that are entirely new. Chief among these are the platforms dedicated to extracting energy, which come in a huge variety of forms and have led to some remarkable island-building technologies.

The first offshore oil rig is a contested title, though it seems likely it was in the USA. Louisiana's Caddo Lake had them in 1911 but documents from Mercer County, Ohio record oil wells pumping far out in the waters of Grand Lake St Marys twenty years before that. Today's rigs dwarf their ancestors. Among those rigs whose legs are lowered to the seabed (called jack-up rigs), the largest so far is the Noble Lloyd Noble, with legs that are 214 metres tall. Spar platforms – where the rig floats like a buoy in the water – are anchored to the seabed and designed for deeper waters. After a two-month journey from shipyards in South Korea, carried by the world's largest heavy-transportation vessel, the main platform of Aasta Hansteen – the biggest spar platform ever built – arrived in the Norway gas fields in 2018. Norway also hosts Troll A, a 'Condeep' (concrete deep) gas platform. The tallest and heaviest man-made object ever moved on Earth's surface, Troll A stands 472 metres high, dwarfing the 381 metres of the Empire State Building. In 1996, its platform and substructure were towed over 200 kilometres out to the Troll field, north-west of Bergen.

Another giant is Prelude FLNG, which looks like a red supersized ship. At 488 metres long and 74 metres wide, the 600,000-tonne Prelude FLNG is the largest offshore structure ever built, displacing six times as much water as the largest aircraft carrier. Created to extract and liquefy gas, it combines production and processing on one huge floating site in the seas off north-western Australia.

The twentieth century witnessed ever larger and more ambitious offshore oil islands. Some went well beyond rigs, such as the four THUMS (Texaco, Humble, Union Oil, Mobil and Shell) Islands in San Pedro Bay. They were built in 1965 to house oil rigs but designed to keep visual and noise disturbance to a minimum. This 'aesthetic mitigation' meant building a phoney landscape. There is a waterfall and luxurious buildings illuminated with coloured lights at night, including a hotel known as The Condo, but they are all fakes designed to conceal the drilling rigs.

The shallow waters of the Arabian Gulf have numerous new islands where oil workers and equipment are located. In 2010 came the completion of twenty-seven drilling islands in Saudi Arabia's Manifa oil field, all connected by a 41-kilometre causeway. The Canadian and Alaskan Arctic has seen some of the most innovative oil and gas islands. The Beaufort Sea has hosted a dozen or so 'sacrificial beach islands', which are made up of beach debris, as well as gravel islands, 'rubble spray islands' and concrete and steel 'caisson islands'. The diversity of island-building technologies in these cold waters is unrivalled. Many are huge. Endicott Island, which is 4 kilometres off the Alaskan coast and covers 18 hectares, is made up of two gravel islands linked to the shore by a causeway. Another technology is 'spray-ice islands', which are among the cheapest to build. One example, built in 1989 by ExxonMobil, is Nipterk P-32. The creation of spray-ice islands in a sub-zero climate starts with hosing water high into the air. The water freezes before it hits the ground, and builds up on the sea ice. In shallow waters, after many days of continuous spraying, the sea ice is weighed down to the ocean floor. The hoses remain on until a roundish island is formed. Completed in fifty-three days, Nipterk was soon able to support a rig as well as service and housing structures.

Offshore wind turbines have no visible platform so it's hard to recognize them as islands. This form of energy production remains expensive and they have not yet achieved a global roll-out. They are only common in north-west Europe, which is also where most of them are built. As of late 2018, the Walney Extension off the north-west coast of England was the world's largest offshore wind farm, followed by the London Array in the Thames estuary. In deeper waters, 2017 saw the completion of the first floating wind farm, Hywind Scotland, 25 kilometres off the Aberdeenshire coast. Although not designed to be occupied, engineers can sometimes be stranded on turbines by high seas, so many have basic food supplies and sleeping bags. Larger offshore wind farms also have separate service and accommodation platforms as well as transformer substations, where the power generated by turbines is collected and sent on to the mainland.

Artificial islands: creation and destruction

The small scale of ancient artificial islands meant they only had a minor impact on the environment. Today's age of islands is different. Many of the new residential and leisure islands like to boast about their ecological credentials. This is mostly 'green-wash', a patina of eco-verbiage designed to convince the incurious. Nearly all modern artificial islands have a deleterious impact on the environment.

On the first island I am headed to, Flevopolder, there are signs that a more careful and considerate model is possible. However, this hopeful message must be set against a sober backdrop. There are four main aspects to the problem: first, the ongoing resources required to maintain an artificial island; second, the knock-on impacts on coasts (new islands change local patterns of deposition and erosion and can lead to rivers silting up and beaches washing away); third, the consequences of dredging on marine life; and finally, the new international trade in sand. Sand is a key resource for the modern world: it is often used to build up islands and you need it to make concrete, the production of which has increased more than thirtyfold since 1950. In just three years, between 2011 and 2013, China used more concrete than the US did in the entire twentieth century. The demand for sand is leading to the destruction of beaches, dunes and seabeds across the world. Two dozen sandy islands have already been completely dug away in Indonesia and 2000 others are at risk of disappearing. Sand smuggling is illegal but today it is a large-scale and lucrative trade.

What happens when you dredge or hoover up the seabed? A lot of islands are built on coral reefs and in places where marine ecosystems have been evolving for millennia. These ecosystems do not pop back up again. New islands often feature inlets and semi-lagoon features where the water is slow-moving, shallow, warm and salty. It may look pretty but this kind of water is oxygen-poor and supports little life. All too often artificial islands are dead zones. Trying to make them live again is hard work. This can be seen at Japan's Kansai International Airport, which opened in 1994. The building of this airport island destroyed the rich seaweeds that were once harvested in the area. The task of trying to replace them has been expensive and uncertain. Gentle slopes have had to be built underwater and seaweeds planted but it remains uncertain if they will survive.

FLEVOPOLDER, THE NETHERLANDS

The world's largest artificial island is going wild: an eco-archipelago has sprouted off a northern shore and its nature park is being reverse-engineered to become primal and untamed. Flevopolder is the place to go to see a new green chapter in the story of the age of islands.

That was not my first impression. Flevopolder is mostly a clipped grid of pancake-flat arable fields and quiet residential streets, and it is huge: at 970 square kilometres it is sixteen times the size of Manhattan – another island first colonized by the Dutch. But, sharing a cup of tea at Ans and Bas's kitchen table, I find it easy to see why they moved here from Amsterdam. The morning sunshine gilds the handcrafted details of their beautiful, self-built home, pouring through tall windows that open out onto acres of woodland, ponds and gardens.

I can't help it: a wave of jealousy is rolling my way. It is with just a bit too much eagerness that I interrupt the polite flow of conversation to ask if they are not worried living 5 metres below sea level. They share a smile before assuring their nervous guest that the island's pumping stations make Flevopolder even safer from flooding than other places in the Netherlands. I'm more likely to end up underwater back in England.

Flevopolder is not a landscape at risk of inundation but it stages another drama. It was declared finished in 1968 but its journey is far from over. Today it is on the frontline of the battle to reimagine what 'artificial land' means and who artificial islands are for. It's a compelling story. Environmental diversity, along with wild horses and wild cattle, is being 'brought back' here.

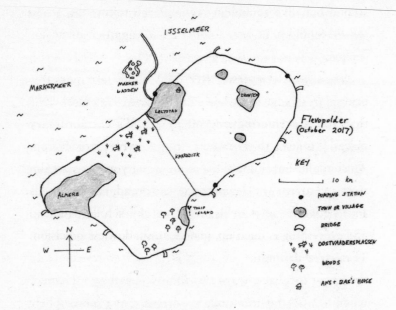

That is, 'back' to a fabricated place, to somewhere that used to be deep below choppy, cold water.

The attempt to rewild Flevopolder is an act of remorse and nostalgia but also courage. This is a landscape that has been won from the sea – shaped, farmed and populated. The Dutch could just sit back and congratulate themselves on the results of their hard work. But they don't. Instead 'artificial islands' are a source of anxious debate about the point and possibilities of human intervention. Today nature is being invited in – a long-lost friend, though perhaps hard to recognize. It's a modern parable for a modern paradox. Industrialism and environmentalism are conflicting yet intertwined ideologies; the first creates the need for the second but, like squabbling twins, each rejects the other. On Flevopolder, at last, some kind of rapprochement has started.

When you first drive over one of the unremarkable bridges from what islanders call 'old land', it's not obvious that you are entering one of the world's extraordinary places. It is just a thirty-minute ride out to Flevopolder from Amsterdam. Leaving pitiless motorways, you cross a calm river and arrive in a geometric space: broad, straight roads slice between green fields. A milky sky scrolls overhead and everything is lined up, just so, in shadowless precision. There is a dreamlike smoothness, a lack of resistance to the landscape. Even its own artifice washes away; the mind rebels against the unsettling reality that every horizon here is a human creation.

My little blue hire car is headed north-east and I soon find myself amid the spacious residential lots of suburban Lelystad, the town named after the creator of Flevopolder, Cornelis Lely. An abrupt left turn and my wheels are grumbling up an overgrown driveway. I'm hoping to find the room I've rented in what looks, from the outside, like a round water tower. I learn later that it's a folly – one of Ans and Bas's creations. Unpretentiously bohemian and wryly humorous, they are bemused by my enthusiasm for their adopted home. Explaining that selling their small flat in the capital bought them a hectare of land, Bas tells me that 'No one wanted to live here.' He pauses, evidently unsure if this attitude has changed: 'If you drive around it's not exactly exciting.' As soon as they were grown, their two children fled back to the city.

'I was crying the first three weeks,' remembers Ans, recalling the young couple's first days here back in 1994. But the regrets have washed away, replaced by a sense of promise. It is with conviction that she tells me, 'There are lots of possibilities here. If you have an idea there are possibilities to create it.'

I've already started to realize that an appreciation of Flevoland's 'possibilities' demands that I learn some more about its past. So next morning, I'm one of the first through the doors at Lelystad's *Nieuw Land* museum. On the ground out front, there's a bright blue rectangle representing the sea, out of which rises a great white head with a life-size Cornelis Lely standing on top, punching the air. As I'm admiring this homage to 'poldering', a black van trundles onto the sea. A ladder is erected and the unblinking white giant is thoroughly cleansed.

Representing Lely's achievement in sculptural form was never going to be easy. The museum is more to my taste: it is stuffed with maps of polders. I am, inevitably, reminded of my earliest days of studying geography. 'The polders' was a staple of post-war geography lessons, at least in Britain. A running joke among my cheeky classmates was the sing-song refrain 'Are we going to do polders, miss?' My love affair with coloured pencils had to remain a secret passion.

Filling in the Netherlands' 'new land' in careful shades – naming the urban areas and the agricultural produce – felt like a kind of magic: conjuring something vast in so tiny a space that it could be closed away in an exercise book. My father, who keeps everything, recently presented me with his own geography exercise book from 1945. In it is a map labelled 'HOLLAND' and its 'POLDER CROPS: WHEAT, BARLEY, RYE, FLAX, SUGAR BEET, POTATOES', and the 'ZUIDERZEE (BEING RECLAIMED)'. Little geographers have been drawing polders for generations and learning their definition (reclaimed and drained low-lying land). The technology to make them is not complicated. What matters is the long-term commitment. The pumps and sluices need to be kept in order; otherwise the water table will rise and you'll be left with a very large bathtub.

Following a major flood in 1916, which killed fifty-one people, Cornelis Lely's solution was accepted: dam and reclaim the Zuiderzee (the huge sea inlet that once scooped out the heart of the Netherlands). It was accepted, in part, because polder-building was nothing new. The oldest ones in the Netherlands date back to the fourteenth century and there are an estimated 3000 across the country. The old Dutch proverb 'the world was created by God, but the Netherlands was created by the Dutch' is no idle boast. Draining the Zuiderzee took a known technology to a new level. These polders would not only create new farmland and living space but also act as protection barriers, preventing flooding across a huge swathe of the country. After a dam was built across the top of the Zuiderzee, the first polder, Wieringermeerpolder, was completed in 1930. It was deliberately inundated by the retreating Nazis in 1945 but reclaimed soon after. Then came Noordoostpolder, which was completed in 1942 and stretched out to gobble up the old fishing islands of Urk and Schokland. Their ghostly shores – curving, wayward – can still be traced in the midst of the surrounding geometry. Both these polders were joined to the mainland. Existing rivers flowed straight over them, causing problems of subsidence. Flevopolder was created as an island to avoid this problem. It has a dividing dyke in the middle, the Knardijk, designed to keep one half safe if the other is flooded. But there has been no flooding. When it comes to building new land, the Dutch know what they are doing.

The expertise of Dutch engineers and designers is behind new islands from Panama to the UEA to the Maldives. Chinese firms are now also acknowledged experts and can undercut on price, but the Dutch remain the respected masters. The specialism of poldering – creating land below sea level – has also gone global. Polder-making has been taken up by numerous countries with flood-prone coasts. You can find polders on the coasts of Britain, Germany, Poland and much further afield. In the 1960s 123 polders – most in the form of new islands – were built in flood-prone Bangladesh. Without them, the regular flooding the country still endures would be even worse. The Netherlands and Bangladesh are both low-lying, river-delta nations and in many ways topographic cousins. Unfortunately, over the past decade, Bangladesh's polders have run into trouble. Their walls are crumbling and the new islands are turning into water-filled rings. Water engineers in Bangladesh now argue that the solution is to work *with* the water: rather than simply keeping it out, we should let it come and go. This new approach involves controlled river flooding: managing the river's flow in, across then out of the polder.

Controlled flooding is catching on in Europe too. In the Netherlands, the 'Room for the River' programme is seeing dykes lowered and some polders, such as Overdiepse Polder (an artificial island that sits in a river near Rotterdam), being cleared of farmland to become a 'spillway'. 'Going with the flow' is the new catchphrase. It's a technical solution to a practical challenge but it mirrors a wider cultural shift. The Dutch want nature back.

Ans and Bas answer my polder questions dutifully but the subject that excites them is the ugliness of the unnatural, modern world. The new towns, Lelystad included, 'turned out really ugly,' says Bas. It dawns on me that their beautiful home – 'we wanted a house that looks older than it is, with nice old tiles' – is another island, a green retreat from a remorselessly industrial world.

It's a viewpoint and an aspiration shared by many. I hear it again a few days later as I'm sitting in a ramshackle basement that is stuffed with eccentrically crafted objects. It is owned by a large, shaggy guy called Ruud, another big-hearted Dutchman with a room to rent, this time in the medieval centre of Haarlem. He is damning about Flevopolder: 'It's got no soul, just money,' he growls before bemoaning the fact that the Netherlands no longer has any real countryside, that it is effectively one big city. He tells me about the time he visited his corporate boss in Lelystad, a town he loathes: 'He wanted to show off his things in this horrible place: I've got this and this and this.' It was all too much for Ruud, who soon fled the corporate world for a wilder, less predictable life.

In every country it's easy to find people chewing over the unappealing nature of modern planning and either finding sustenance in the resultant melancholic cud or actively searching for something better. Today the Dutch are also leading the way in the ecocentric redesign of new islands.

Back in Flevopolder, it is time to leave the *Nieuw Land* museum, especially now it's getting full of skittering schoolchildren, hunting in packs for buttons to push. Outside the weather has closed in: a steady rain smears the windscreen as I drive north, along the 30-kilometre dyke road that cleaves what was once the Zuiderzee. By the time I get a quarter of the way up, at what I hoped would be a panorama of eco-island splendour, it's pelting down. On the right is the IJsselmeer, on the left the Markermeer, two vast freshwater lakes. Lely planned a polder for the Markermeer but his scheme was finally shelved in 2003. A new, greener mood had taken hold; the prospect of another century of creating land exclusively for human use had become unappealing, especially as it was learned that the Markermeer was silting up, its sea creatures suffocating and bird life disappearing. After parking I brave the bleak drizzle, and the new solution for the Markermeer is just visible in the mist. It's incredibly bold. Wiping raindrops from my glasses, I can make out the sinuous black lines of new dam walls. They loop and roll over much of the horizon. This is a new eco-archipelago. Not an exercise in geometry but a twisting labyrinth, it's easy to envision as a maze of islands, creeks and hidden places.

The new 10,000-hectare eco-archipelago, which officially opened in September 2018, is called the Marker Wadden. The first island was begun in 2016, made by mud dredged up from the lake's silty reaches, thus deepening and clearing the lake at the same time as providing a natural habitat for wildlife. When settled into place, the islands will be covered in reed beds and low dunes and surrounded by rock break-waters to stop the whole thing being washed away. A little harbour will allow hiking day trips for nature lovers but that will be as far as the human presence goes. The results are not entirely predictable; that also is part of the design. This is planning with nature; therefore nature will 'get a say' in what happens and what the Marker Wadden looks like. Project manager Ruud Cuperus, talking to *Het Parool* newspaper, is pleased that 'herring, smelt, glass eel and anchovies are now swimming again' in the surrounding waters and he looks forward to the arrival of more bird life, including exotics like wintering flamingos and spoonbill. But he is candid that this is a journey into the unknown: 'You can pull all kinds of levers but the outcomes are not predictable.'

The Marker Wadden is not the only green archipelago being built. There are others in the Markermeer and off the other coasts of Flevopolder, including islands shaped in the form of a tulip. Sculpting shapes for aerial photo opportunities is a common weakness of island designers. A more convincing green archipelago was built adjacent to Amsterdam at the start of the century. IJburg is all about sustainable human habitation; the first residents of this ten-island complex moved in in 2002. Designed for minimal carbon emissions, with communal roof gardens and plenty of parks and trees, IJburg has proved popular – and not only because it has a mixture of prices, with plenty of low-cost rentals. Increasingly people want to live alongside and with nature; the straight lines and efficient spaces of Lely's day are no longer seen as exciting but as unsustainable and boring. There is a seismic shift underway in how the Dutch think about the purpose and consequences of 'new land'.

The most controversial outcome of this shift is the re-wilding of 56 square kilometres of northern Flevopolder. I've been saving the trip to Oostvaardersplassen until my last day on the island, though I know it is not an experience to be rushed. After all, this is Europe's best-known example of rewilding: not a landscape designed for human pleasures – including the pleasure of staring at animals – but a place to be left alone.

That said, there is a visitor centre with a cafe and gift shop, from which point you can wander off across umpteen tracks. I opt for one that is heading seaward, with just the sound of the wind in the tall reeds for company. It's so flat and wide that distances are hard to gauge, and it's difficult to imagine that this empty quarter was once earmarked for industry or that it is now a battleground between those happy to see its various species live and die here and animal rights campaigners who charge it with being a cruel and unnatural experiment where large herbivores starve to death. For some, claiming this place as 'primal' is a piece of artifice, a conceit for which the animals trapped here (there are currently no green corridors for them to move off the site) are the innocent victims. The horses and cows introduced here in the mid 1980s were selected for their archaic qualities. The Heck cattle are a hardy breed, first bred in the 1920s to resemble the extinct aurochs, an ancient bull-like species that once roamed Europe and Asia. The horses are Konik ponies, which look like another extinct species, the Eurasian wild horse, the last of which died in 1909. Along with red deer, these animals spend all year out in the open and, in warm years, their numbers have outstripped the natural feeding capacity of the reserve. Since they have no predators, rangers have stepped in to cull the animals. In 2018, local politicians decided that the number of large herbivores should be capped at 1500. They were responding to a rewilding backlash that was unleashed when photos of starving beasts circulated on social media. Protesters went so far as to compare Oostvaardersplassen to Auschwitz. Groups of them began tossing bales of hay over the fence that surrounds the reserve. Campaigners like behavioural biologist Patrick van Veen are adamant that Oostvaardersplassen is a 'failed experiment' tarred by 'machismo and deceit'.

Yet, heading further along this increasingly marshy path, it occurs to me that if Oostvaardersplassen is a failed experiment then Flevopolder is an even bigger one. After all, creating a geometrical island where unproductive nature is banished is a pretty extreme idea. You can't judge either experiment in isolation from the other. Oostvaardersplassen is one element of a shift away from just thinking about the landscape and the planet in terms of humans and their immediate needs. It's not an easy thing to do; our species has come to define itself as beyond or above nature. The bonds have broken and sticking them back together will often look fake and feel clumsy. But it's the right thing to do.

After hours of contemplative wandering, I'm back in my small blue car, ready to head off across the bridge, back to 'old land'. I pass a group of Konik ponies gently cropping a wild meadow. They must be used to sightseers as they barely stir. There is a line of parked cars ranked nearby, with cameras poking out of every window. There used to be not much to see in Flevopolder but now there is. Wild animals are exciting, alive and effortlessly beautiful in a way the human world isn't. We are drawn to them, as to every tiny plant or buzzing bee, irresistibly and inevitably. Our love of other forms of life is a self-preserving impulse: bringing nature 'back' into the picture is not just pleasing; it is necessary for our survival and our sanity.

THE WORLD, DUBAI

It was raining in Dubai as I stepped onto Lebanon. Later that day journalists heralded the stormy weather as proof of the efficacy of intensive cloud-seeding. Artificial rain on an artificial island, falling on one of only two completed islands of the 300 that make up 'The World'. The other is in the Greenland group, reportedly a gift by Dubai's ruler, Sheikh Mohammed bin Rashid Al Maktoum, to Formula One racing driver Michael Schumacher.

When seen from the window of a plane The World, sited a few kilometres off Dubai's downtown shoreline, is a plausible world map. The continents are all there, though things go a bit wonky at the top, bottom and around Australia. Each continent is made up of roundish sand islands. A lot of them are assigned to particular countries and they sit in roughly the right place. So, for example, Egypt is above Sudan, which is next to Eritrea and Chad. Some of the biggest countries, like Russia, are broken up into islands that represent cities or regions; so there is a Moscow island and islands for Omsk and Siberia.

The only way to get to The World is by boat and, so far, there are no bridges between the islands, though many of the channels between them are narrow and shallow. If I wasn't wet enough already, I might be tempted to wade over to Palestine.

The original vision of The World, launched by Sheikh Mohammed in 2003, was that the islands were to become playthings and unique retreats for the super-wealthy. Today,

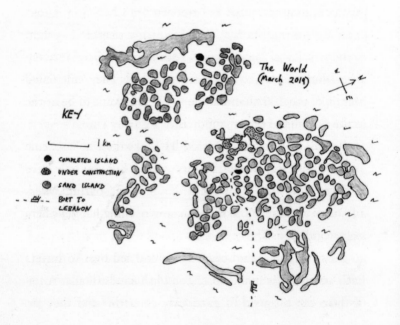

however, it's just me and a family of three from Glasgow on a day-return that costs £40. We are met on Lebanon's pier by its Indian manager, who is smiling steadfastly as the water pools on the little white tray he carries bearing four cups of pineapple juice. Apart from him and a few other staff, we're the only ones here.

Nakheel, the state-owned firm that built The World, initially sent out invitations every month to 'Own the World' to fifty rich, high-profile potential buyers. Karl Lagerfeld had plans for a 'fashion island'. It was said that Brad Pitt and Angelina Jolie had bought Ethiopia for their children. The photo opportunities and the showmanship were irresistible. Richard Branson posed on Britain in a Union Jack suit next to a British phone box. Turkey's MNG Holdings bought Turkey; China's Zhongzhou International bought Shanghai. Back then Nakheel's marketing people depicted The World as *facing away* from Dubai. They calculated that, like other aristocrats, the super-rich want to be at the centre of things but remain invisible.

It's easy to find critics who will tell you the whole scheme is bonkers. It is, undeniably, outrageous. But you only have to turn round and look back at the twisting, soaring towers of Dubai to understand that this sort of thing happens here on a regular basis.

Most of The World islands were turned over to buyers in 2008. Nakheel was basking in the success of another of its projects: the world's most famous artificial island, the

Palm Jumeirah. The fronds and trunk of the Palm Jumeirah stretch out over 5 kilometres and, if you include the tourists staying there, today it has a population of about 75,000. Repeating that success has not proved easy. Just after completion of The World, the worldwide financial crash rolled into town. The crash messed up building timelines for The World and a bunch of other islands Nakheel was constructing or planning. Another huge palm, Palm Deira, was downsized and lopped off into the family-friendly resorts of the Deira Islands, which are well underway. The giant sister of Jumeirah, Palm Jebel Ali, has been piled into shape but further work was put on hold and its heralded halo of islands – shaped into an Arabic poem written by Sheikh Mohammed – quietly forgotten. In hindsight, the Sheikh's would-be island-words have an ironic subtext:

Take wisdom only from the wise,
Not everyone who rides a horse is a jockey.
It takes a man of vision to write on water,
Great men rise to great challenges.

Another casualty was The Universe, a delirious, cosmic fantasy that would have wrapped The World in bands of islands forming the shape of the Milky Way and the solar system.

The World may have found a saviour in Josef Kleindienst. Once a policeman and member of Austria's right-wing

Freedom Party, he is now one of Dubai's big property developers. Arriving at the western shore of Lebanon, and looking out over the empty shoreline of what I'm told by another Indian staff member is Syria, the skyline is animated with multiple cranes slotting together the Kleindienst Group's 'Heart of Europe'. (The young member of staff also expressed irritation that the island of India was smaller than Pakistan: 'That is very wrong.') Six territories are being built – Sweden, Germany, Main Europe, Switzerland, the heart-shaped 'honeymoon island' of St Petersburg, and The Floating Venice – in order to form an interconnected upmarket resort. The golden domes of Sweden's beach mansions and Germany's modernist executive villas are nearly complete. In addition, there will be floating individual three-storey 'Seahorse Villas', some of which have already been built and which feature underwater windows.

The target audience for The World has broadened. The scale of the rising blocks shows that reclusive exhibitionists with deep pockets are no longer the key market; these are not solitary domiciles for tycoons but hotels, apartments and shops. The Heart of Europe is being built to accommodate up to 16,000 people and its promotional spiel promises 'European retail coupled with food and beverage concepts' as well as sheer spectacle, such as 'ground-breaking climate control technology which will convert narrow cobbled streets and picturesque plazas into a beautiful winter wonderland!' Again the outlandish nature of this vision needs to be placed in context: Dubai's Mall of the Emirates already has a huge real-snow ski slope and its many supersized 'shopping experiences' vie to outdo each other in showmanship.

Both Dubai and Kleindienst thrive in the meeting point between social conservatism and huge ambition. It is a fertile terrain for hyper-consumerism and artificial islands. Environmental, political and welfare concerns, which would derail such schemes in Europe, barely get a look in here. Despite a much-proclaimed shift to renewables – especially solar power – creating air-conditioned, shopaholic lifestyles in the desert eats up a lot of resources. The United Arab Emirates, of which Dubai is part, is run by a patriarchal fiefdom and its shining towers and motorways are built by an underclass of South Asian men who work in incredible heat. A local building engineer explained to me that it is illegal for workers to be out in temperatures beyond 50 degrees Celsius, 'but it is funny how the thermometer gets stuck,' he grinned at me, 'gets broke at 49.8 or 49.9, you know. It's not good but that happens.'

The environmental costs of Dubai's new islands mostly concern the huge amounts of materials and energy required to build and sustain them as hot spots of high-income, air-conditioned mobility and shopping. It's not all bad news, though: in terms of marine life, the islands may have a positive impact. Along with the protective reefs that surround them, they provide a habitat for corals, fish and other sealife that would otherwise not find a home in the Arabian

Gulf's shallow, sandy waters. Artificial reefs – made of everything from boulders to sunk trains – have been shown to encourage biodiversity in many parts of the world so it is not surprising the same thing happens here. Nakheel's most audacious reef loops round the northern reaches of The World. Over a thousand coral-covered boulders that were under threat at a port site further up the coast were towed 14 kilometres underwater. Nearly all the coral survived, and this section of The World reef now attracts divers.

I pick my way around the unoccupied beach furniture on Lebanon. Occasionally an employee will brush a few damp leaves from the sand or look out of a window from the empty restaurant. When the rain comes down harder, I head inside and become the restaurant's only customer. Delivering my sandwich and chips, the Kenyan waiter tells me about his plans. He works twelve hours a day, sending money back home to buy a supermarket. The idea is that his wife will run the supermarket and he will buy a poultry farm. That's his dream and it should be possible, he says, in three years. The manager and beach-sweepers have similar ambitions. To me Lebanon seems like a place dropped out of a scene from a Samuel Beckett play – all silences, ennui and emptiness. But I suspect this says more about me than the island: I can afford to indulge my boredom. By contrast, many of the people who build, staff and get customers to and from these resorts don't have time for feelings of 'ennui'; nor do they see themselves as victims but rather as grafters, even as would-be entrepreneurs.

With 300 play-sized kingdoms being traded to the world's richest individuals, it's no surprise that all sorts of stories – many of them true – swirl around The World. Another project that did not come to pass was Opulence Holdings' scheme for Somalia; the idea was to shape it into a seahorse and build luxury houses where residents could hit golf balls into the sea from their balconies. The owner of Ireland, John O'Dolan, had plans for a replica of the Giant's Causeway but took his own life when his debts became insurmountable. The purchaser of Britain ended up in prison for bouncing cheques. Many owners, like Baron Jean van Gysel, the Belgian hotel owner who bought Greece, are biding their time. When he bought Greece, van Gysel said that his first act would be to run a metal band around the island to protect it from erosion. As far as I know, the metal band hasn't materialized, but it throws up the question of how other owners are going to protect their islands. Or, indeed, to supply them with water, power and waste disposal. The initial plan was for fresh water and electricity to be laid on, piped from the mainland out to The World. However, for the time being, owners have been left to their own devices – which, in the short term, means diesel generators and shipping stuff back and forth.

Despite the challenges, the allure of The World is still drawing in investors. Two big green-lit proposals are for a resort called OQYANA on the fourteen islands that make up Australia and New Zealand, and for a low-rise resort village on twenty islands in the North America group. One of the more recent investors was Hollywood actress Lindsay Lohan, who is designing her own island and giving it the name 'Lohan Island'. She told *Emirates Woman* that the island will feature a 'luxury hotel, Michelin-worthy restaurant, idyllic waterfront pool and plenty of leisure activities'. Lohan also revealed that she has bought Lebanon and plans to revamp it as a 'luxury getaway'. From where I'm standing, it looks like a 'luxury getaway' already, albeit a rather damp one. But 'luxury' is a restless animal, forever feasting but never satisfied.

It seems inevitable that Dubai's luxury skyline will colonize The World and continuously shape and reshape its islands. On the other side of Lebanon to the Heart of Europe you can see the virgin lands of Palestine, Jordan and Saudi Arabia and, beyond that, the massing forces of the future, including the hazy pinnacle of the world's tallest building, the Burj Khalifa.

After a few hours on Lebanon everyone, staff included, took the boat back to Dubai. I want to see what it is like actually living on one of Dubai's islands so I've rented a room on the Palm in the home of a young expat couple called Reena and Ryan. Pointing out the shops across the busy road that runs up the spine of the Palm, Reena jokes that I might need to get a taxi. Later, edging my way along a broken and vestigial pavement, I realize she was not joking. Reena and Ryan have a three-year-old daughter, a happy chatterbox who absorbs most of their energies. She spends much of the day in the front room with views over the bright, glinting waters and the low-rise mansions that occupy the Palm's fronds. The little girl rolls restlessly between a large television that plays endless cartoons and another screen that scrolls out educational games. Both parents worry about her: she's boxed up when she should be playing and running outside. It's way too hot out there and, apart from some manicured green strips, there is nowhere to go. 'I do feel that she is missing out on something that I took for granted,' says Ryan. 'We do what we can; it is a big thing that concerns me.'

But they are not planning on moving. They tell me the compensations outweigh the costs. It's not just the money they can earn here and the zero-rate income tax, it's also the fact that Dubai is safe and efficient; there is very little crime and almost no rubbish on the streets. Ryan and Reena don't even bother to lock their front door. They've travelled widely and been to plenty of countries, including the UK and USA, and have no appetite for what they saw: insecurity, inefficiency and dirt. It's an ironic truth that, although Dubai panders to the rich, people with great wealth are insulated wherever they live: they don't need Dubai. It's ordinary people, like Ryan and Reena, who endure the hard edges of anxious, neglected places and it is they, above all, who value this modern wonderland in the desert.

Dubai is used to put-downs. Even the *Rough Guide to Dubai* is condescending about the Palm. 'Disappointingly botched,' it says, with 'densely packed Legoland villas strung out along the waterside "fronds"'. A visiting journalist for the *Guardian* describes 'rows of McMansions looking across at each other between thin strips of stagnant water'. I suspect there is an element of resentment in such dismissals. Love it or loathe it, the architecture here – mansions included – is often bespoke and frequently daring. Countries, like the UAE and China, that were once poor are poor no longer. They have seized control of the image of the modern city. Critics in the West, who thought they had the copyright on modernity, are becoming dimly aware that they are now outside the loop.

Also envious of Dubai's success, other Gulf States have been creating their own artificial leisure and residential islands. Qatar's Pearl is nearly finished; its developers promise homes for 6000 and 'a warm, welcoming community whose residents are seeking an urbane and vibrant lifestyle'. In Bahrain there are the residential communities of Durrat Al Bahrain, shaped like petals floating in the sea, as well as Northern City and the Amwaj Islands. Kuwait's Green Island led the way when it was opened in 1988, though today it is somewhat overlooked – especially now a complex of finger islands called Sabah Al Ahmad Sea City is nearing completion and is already being called the Venice of the Desert.

How long will these islands last? In Nakheel's HQ I was told that the Palm islands are built to withstand a predicted sea-level rise of 0.5 metres and that they are 4 metres clear of the water. The first figure seems conservative: most scientists predict a larger rise. And from what I could see, the islands of The World are nothing like 4 metres clear of the sea. Moreover, with 85 per cent of the population of the UAE living in vulnerable coastal areas, the threat of sea-level rise on the islands cannot be separated from the possible inundation of the urban shoreline they are plugged into. Another worry is the temperature. Climate change is predicted to make the Gulf States even hotter. It is already too hot to safely be outside for much of the year, so the question of how habitable Dubai will be in fifty or a hundred years' time is a real one.

Delegations from around the world come to Nakheel's offices to learn and copy. 'We have a lot of governments coming to learn from us,' they told me: 'from China, South Korea and now from Africa as well. We get a lot of people wanting to emulate what we have done.' The so-called 'Oriental Dubai' is called Phoenix Island in China. The axis of island-building is shifting eastwards; as we shall see in the next chapter, though, many of them are not about leisure and pleasure but more hard-nosed modern needs.

CHEK LAP KOK, AIRPORT ISLAND, HONG KONG

Having obliterated a place, modern planners like to decorate their new creation with street names and signposts that sweetly memorialize what was there before. 'Oak Tree Grove' has neither oak tree nor grove; 'Green Acres' is a jigsaw of tarmac and brick. I'm reminded of this perverse practice as I squint up at a bright white sign announcing 'Scenic Drive'. All around are shrieking roads, high wire fences and the geometric assemblages of airport hangars and terminals. I'm at the start of a day trip to an artificial island dedicated to Hong Kong International Airport. It is called Chek Lap Kok and, at 1248 hectares, it is almost twice as big as Gibraltar.

The mouth of the Pearl River is spangled with mountainous islands. On some of the gentler slopes skyscrapers jostle for every spare inch but on others subtropical forest lies undisturbed. Hong Kong has both; it is a gloriously vertiginous and idiosyncratic city-state that has an uneasy relationship with the Chinese mainland. In 1989, a few days after the massacre of pro-democracy protesters in Beijing's Tiananmen Square, a new airport was proposed by David Wilson, the British Governor of Hong Kong, which was still a British dependent territory. The new scheme was widely seen as a bid to bolster the confidence of Hongkongers fearful for their future under Chinese rule. Compounding this impression, Wilson announced a draft Bill of Rights a few weeks later. For their part, the Chinese were irked by Britain's high-handedness in piling up a tab they had no intention of paying. President Jiang Zemin responded angrily: 'You invite the guests, but I pay the bill!'

Given its size, Chek Lap Kok was built remarkably quickly: commencing in 1992, it rolled off the production line in July 1998. It was crafted by flattening, expanding and connecting four existing, natural islands (including the old Chek Lap Kok). A sizeable chunk of the world's commercial dredging fleet was involved. They dug sand from the seabed and sprayed it in great arcs until a smooth and even platform emerged. At the same time, 34 kilometres of roads, tunnels and bridges were built, along with an express rail link. In all contractors relocated 238 million cubic metres of material, which is comparable to moving a small mountain.

The new island saw the destruction of 50 hectares of mangrove and the mass transplantation of both people and endangered fauna. The old Chek Lap Kok was about 3 square kilometres in size, an ancient and hilly island that had once been a favoured hideaway for pirates. The brigands had long gone, leaving a mining and fishing community, a Qing temple and a shrine dedicated to the sea goddess. The villagers were relocated to a set of anonymous blocks called Chek Lap Kok New Village on a neighbouring island, complete with a rebuilt temple. The human population accepted their marching orders with meek endurance. The shy and secretive Romer's tree frog caused more fuss. Discovered in 1952, this fingernail-sized, dowdy amphibian is unique to Hong Kong and its uncertain fate quickly established it as an icon of Hong Kong wildlife, so 230 specimens – about half a bucketful – were transported off the island and installed in safe havens. It is rumoured that a few of the tree frogs cling on in wet patches of their original homeland. If so, they are unlikely residents of one of the world's most expensive airport developments.

For island-spotters, Chek Lap Kok is a magnificent example of an 'infrastructure island', an offshore platform dedicated to activities too polluting, noisy, unsightly and hazardous to be tolerated elsewhere. Japan led the way. It has five offshore airports, including Kansai International Airport. Completed in 1994, and situated over 5 kilometres from shore, Kansai was the world's first wholly artificial airport island. This type of island usually has a penumbra, a rim of land left empty that forms a barrier between it and the ordinary world. In Chek Lap Kok's case, this empty zone was designated as 'noise-sensitive land' – somewhere the scream of aircraft is way above acceptable levels.

It is around this benighted edge that I'm taking my afternoon stroll. A few metres from the 'Scenic Drive' sign, I'm relieved to see a walkers' fingerpost. It beckons me with unlikely promises, pointing the way to 'Scenic Hill', 'Ancient Kiln' and a 'Historical Garden'. A fierce twelve o'clock sun is stinging my skin and these nostalgic destinations suddenly appear less attractive than the cool, deep shadow pooling under the Hong Kong–Macau bridge, which shoots over one side of Chek Lap Kok on fat concrete stilts.

On the day of my visit, in April 2018, the world's longest sea bridge is not quite complete and remains silent, a slumbering colossus. It rides out seawards, 55 kilometres long, linking China's two mini-states and tying them firmly to the motherland. The bridge has two striking mid-route artificial islands. Resembling sleek ocean liners, they funnel the roadway into a 6.7-kilometre tunnel, allowing unhindered sea passage for the container ships that ply the Pearl River delta.

Chek Lap Kok and the bridge to Macau are mega-projects, heroic in scale and internationally significant. But, resting in the bridge's shade, I can't help thinking about where Hong Kong is heading and the sense of loss that comes with transformation. Perhaps these are my problems. My attraction to places no one in their right mind would want to go can sometimes feel worryingly masochistic. The 'walking track' indicated by the fingerpost is strewn with building rubble. I've yet to see another soul. Who is all the signage aimed at? Back in the unforgiving sun, I find an answer on a battered information board. In English and Chinese it natters on about the 'Airport Trail' and was presumably put up to instruct schoolchildren dragged here by merciless teachers. 'What is the purpose of the trail?' it demands. Somewhere in a hot recess of my skull I graffiti a response that I'm sure also occurred to a few of those children: 'to remove all hope'.

This reminder of educational duties triggers a more immediate anxiety. I'm in Hong Kong as a staff member on a university field trip with twenty-nine geography students – or is it twenty-eight? – investigating ideas of citizenship. Thankfully the field trip is led by Michael, the young lecturer who is enthusiastic, knowledgeable and has a generous smile that could warm the most homesick heart. He even gets up before everybody else to buy punnets of fresh fruit to supplement the students' breakfasts. It soon becomes obvious that all I have to do is stand next to Michael, not eat all the grapes and occasionally repeat what he has just said. In a tacit admission that this is neither demanding nor useful, Michael suggests I take the afternoon off to pursue my private passion for islands.

Yet the field trip has been an intense experience and I can't help thinking about the encounters it has led to. We've been speaking to pro-democracy activists and going into highly charged spaces. A year later, protests would explode into the streets and be heard around the world but, when we were there, it felt as if overt opposition had been successfully muffled and was disappearing. On our first day, Michael instructed us all to sit down in a circle on the tiled plaza beneath HSBC headquarters. This was one of the key sit-in and tent-city sites for what we, mistakenly, imagined was the high-water mark of Hong Kong protest, Occupy Central, a movement that fed into 2014's Umbrella Revolution, which saw 100,000 protesters take to the streets. Michael knew that the sight of so many people sitting down on this signifi-cant spot would lead to an appearance by the security forces. He also knew they weren't likely to be too scary. They duly arrived. After twenty seconds China's iron fist was made manifest in the form of a portly, diminutive and very polite female security guard who ambled slowly over to inform us that sitting down was not allowed. This clash with totalitarianism created quite a buzz among the students and I basked in the reflected glory of Michael's chutzpah. Later encounters were more telling: activists speaking confidentially and quietly, not wanting to be overheard: 'Is the CCTV on here?' One older man informed me matter-of-factly that 'Hong Kong is disappearing very surely.' From 2019 a new wave of protests began and the subsequent crackdown has fulfilled our interviewees' worst fears. On 30 June 2020, a law was passed by Beijing that means dissent in Hong Kong is now just as dangerous as it is on the Chinese mainland.

We heard other, less expected concerns: about the intrusion of an alien, uncouth Chinese culture ('They are so rude; they push past') and the threat to Cantonese, the regional spoken version of Chinese, which is making way for Mandarin ('It is wrong that the ten-year-old boy, my neighbour, is speaking a different language').

My footsteps begin to connect the dots. Chek Lap Kok and its giant bridge to Macau are at the epicentre of China's plans for the future: here too somewhere distinct and small is being rolled over by powerful forces. Here too I am encountering forlorn attempts to hang on to vestiges of the past while, all around, the ground shifts.

My trail does, as promised, lead to a small archaeological site and a patch of grass labelled 'Historical Garden'. I catch up with a group of five female airport employees, each wielding a bright parasol, one of whom tells me they have driven out from behind the airport's high fences to touch and smell the flowers. At the 'ancient kiln' there are rows of explanation boards that go into detail about how the 'airport retained the hill at the southern tip of Chek Lap Kok Island' where 'furnaces ascribed to Yuen dynasty (AD 1271–1368)' have been unearthed along with 'Neolithic remains' and 'ceramics of Tang and Song dynasties'. The jewel of the historical garden turns out to be a 2-metre-tall, many-toothed 'cutter head' from a 'cutter-suction dredger'. Painted a fetching blue and framed by acacia trees flushed with scarlet flowers, it squats fatly like a malign Buddha. The smiling women pose for pictures in front of this revered object.

For a moment things fall into place and I imagine a kind of balance has been struck between the old and new, yin and yang. The fantasy is soon torn apart. This island is sprouting new limbs and the wide bay it lies within is churning with restless industry. Soon after leaving the comforting aura of the parasol-holding women and trudging reluctantly towards massed cohorts of diggers and construction barriers, I'm immersed in dust and grinding gears. Coughing and leaking sweat, I round a headland and everything stills. In front of me is a panorama of truck-sized pipes; they are scattered across the shore and the shallows like penne fallen from the plate of a boorish pasta-loving giant. Land and water are radically confused: platforms and concrete plinths jut out of the sea for miles around. A little further away, shimmering in the heat, another new island is being birthed. It is a polyp of Chek Lap Kok and connected to it by a thin neck of road. I clamber my way forwards, walk inside the length of one of the pipes, and get as close as I can to the diamond-shaped 150-hectare newbie. This island is dedicated to the single task of processing traffic on the Hong Kong–Macau link and its official and only name is 'Hong Kong Boundary Crossing Facilities'.

The tediously literal place name isn't the only reason Hongkongers aren't paying much attention to what, in most other countries, would be a source of excitement and controversy. So much else is going on, and people here are well used to land being won from the sea. The shoreline across the city has been pushed forward for 150 years and there are other huge schemes afoot that are soaking up media attention. Off Chek Lap Kok's northern shore dozens of barges and dredgers are at work, building 650 more hectares for a third runway. The Hong Kong government's '2030 Plus' plan envisages a new 'infrastructure island' for an incinerator as well as a 1000-hectare urban island with a population of up to 1.1 million. Dubbed East Lantau Metropolis, it will create a new business district as well as platoons of new apartments. In a city where the average apartment costs over eighteen times the average income – and where, consequently, home ownership is an impossible dream – 'new land for new houses' is a popular vision.

But why more islands? They will be hugely costly, eating up at least half the city's fiscal reserves. The new islands are vulnerable in other ways too. Hong Kong is expected to experience a significant rise in sea level as well as an increase in the frequency and intensity of typhoons. Forecast maps show that much of Hong Kong's reclaimed land could be underwater by the end of the century.

Hong Kong's island frenzy is a puzzle. I suspect it isn't just about sensible choices and expert decisions. Even though the islands squeezed out from its 'cutter-suction dredgers' have the romance and visual appeal of a piledriver, they are still amazing. They reflect a preference for the drama of the open sea over the open arms of China.

I've been walking for hours now and I packed almost nothing to bring with me: a single banana and two water bottles. I sit down on a large boulder in the shade of another road bridge and watch some young men precariously perched on one of its piers, a concrete shelf they have paddled across to in a tiny metal boat. One of the youths, wearing only bright yellow shorts, leaps up and starts circling his thin arms, shouting to me or at me. I just stare back, aware something is amiss. What is wrong? The man calmly settles down, suddenly indifferent. Perhaps he was signalling to someone far away. It's easy to misread this place.

Getting the measure of Chek Lap Kok can't be done in its generic airport terminals but it's hard here too. The island seemed more graspable yesterday, when I took the cable car that leaps over its shoulder on the way up to see the world's largest outdoor bronze seated Buddha. The ride affords huge views. You can follow the snaking curve of the Hong Kong–Macau highway out to sea and try to count the myriad vessels piling up land for Chek Lap Kok's third runway. The island is crawling with aeroplanes, like white flies on a corpse. From high above, it looks like a landscape that has been killed and skinned. So why does it hold my attention and draw my eye down to every desiccated shape and shore? Maybe because of its sheer oddness, the way it manages to be both machine-like and magical, somewhere conjured up from nothing.

In the cable car's rattling, glass-bottomed gondola another answer occurred to me. Travelling over a fabricated island built for travellers, it felt like Chek Lap Kok was the epicentre of a global culture in which impermanence and mobility is everywhere and prized beyond anything else. The flip side of this restlessness is a yearning for the lost authenticity of place, and nostalgia for a less relentless and uprooted way of life. The Hong Kong activists I've been listening to are caught up in this same dilemma. They don't just want a vote; they want to hold on to something of value, an identity and a history that is being dug away and flattened out.

Heading away from Chek Lap Kok, back to my hotel, I make my solitary way onto another island then another. I walk across skywalks, up escalators, into subways, never quite sure if I'm ever in contact with solid ground. The ceaseless churn of this shifting city fills each moment with thoughts of departure and destination – to better places and worse.

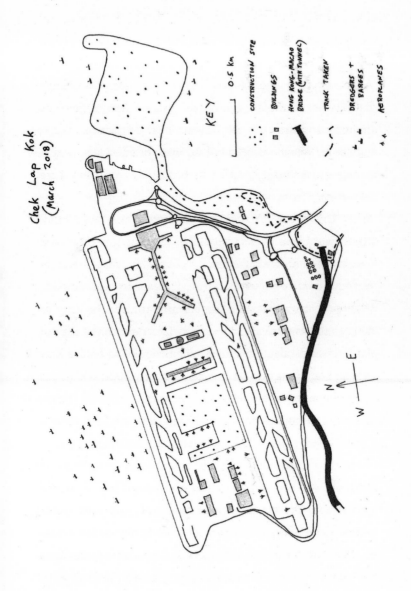

Chek Lap Kok
(March 2018)

KEY

0.5 km

CONSTRUCTION SITE

BUILDINGS

HONG KONG-MACAO
BRIDGE (WITH TUNNEL)

TRACK TAKEN

DREDGERS +
BARGES

AEROPLANES

FIERY CROSS REEF, SOUTH CHINA SEA

Of the seven remote reefs transformed into landing strips, harbours and missile silos by the Chinese military in the South China Sea, Fiery Cross is the most important though not the largest. That distinction goes to Subi Reef, which has been bulked up to nearly 4 square kilometres and is crowded with over 400 buildings. In its natural state Fiery Cross was a ragged coral death-trap for unwary ships. Today it is 2.8 square kilometres in size and China's key forward base in the region. The new island is claimed to have twelve hardened shelters for missile launchers, an early-warning radar and sensor array, hangars to accommodate twenty-eight combat and bomber aircraft, and housing for more than a thousand troops. Fiery Cross also has a runway that is over 3 kilometres long – long enough to land a Xian H-6 jet bomber, which has a range of nearly 6000 kilometres.

Military reclamation of the island began in 2014. 'Before and after' photographs show the conversion of the natural reef – full of colour and enclosing a large pale blue lagoon – into a grey rectangle with a long black airstrip and gaping square jaw. The jaw is the military harbour and is usually dotted with the black teeth of destroyers and other naval vessels. The Chinese claim nearly all of the South China Sea, leaving the other nations that surround it with residual coastal strips. Unsurprisingly, China's claim is hotly contested and the Permanent Court of Arbitration – an international body that tries to settle intergovernmental disputes – has ruled that there is 'no legal basis for China to claim historic rights'. There are many overlapping claims on the numerous islands that make up the Spratlys (of which Fiery Cross Reef is a part). Ownership of Fiery Cross Reef is claimed by the Philippines, where it is called Kagitingan Reef, as well as by Taiwan and Vietnam.

The Chinese call it Yongshu Reef. 'Fiery Cross' remains its international name – one with a suitable ring, given how perilous the current stand-off is in the South China Sea.

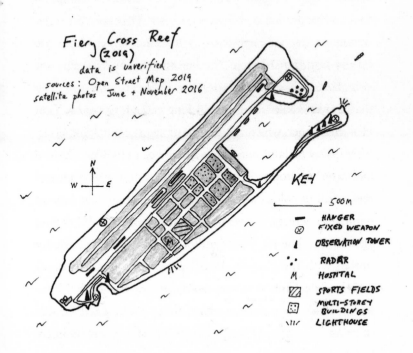

Fiery Cross Reef (2019)
data is unverified
sources: Open Street Map 2019
satellite photos June + November 2016

N
W — E

KEY

|— 500M
— HANGER
⊗ FIXED WEAPON
↟ OBSERVATION TOWER
∴ RADAR
M HOSPITAL
▨ SPORTS FIELDS
▦ MULTI-STOREY BUILDINGS
\ιι⁄ LIGHTHOUSE

The name derives from 31 July 1855, when *Fiery Cross*, an English 'extreme' tea clipper famed for its speed, ran aground here. This part of the South China Sea is called Dangerous Ground, a reference to the number of ship-snagging reefs. With China showing no sign of accommodating its neighbours and regularly seeing off planes and fishing boats that stray near its de facto possessions, this is another newly appropriate place name.

The Chinese military strategy appears to be to establish a forward line that gives them reach over the whole of South East Asia. Taking control of the South China Sea also has economic benefits. Over $5.3 trillion worth of shipping travels through these waters every year. They contain extensive and untapped oil and gas reserves and about 12 per cent of the world's fish catch. The Spratly Islands are the aces in a high-stakes game of geopolitical poker.

The building of new military islands requires huge resources and perseverance. The first task is to find a reef that can provide stable and enduring foundations. China is not the only country that has done this in the South China Sea. Taiwan, Vietnam and Malaysia have transformed the reefs they control in the same way. Visiting them is not easy. Tourist boats from Vietnam and China do go on trips but these are patriotic missions and only open to vetted nationals of those countries. The only military Spratly Island that foreigners can visit is the one controlled by Malaysia. It is called Layang Layang and it looks like all the others: a rectangular landing strip. However, Layang Layang also has a 'diving hotel' that provides package holidays for those who want to explore what remains of the corals that surround the island.

The closest I've come to Fiery Cross Reef is looking out over the South China Sea from the palm-fringed city of Sanya, China's most southerly city. I have a local tourist map with an inset of the whole of the South China Sea. It shows a thick dotted line marking China's territorial argument and which delivers almost the entire sea to China, with little red squiggles in the centre representing the Spratlys and Paracel islands, all claimed by China.

It may seem odd that the South China Sea should be included in a local tourist map. The rest of the map is concerned with pointing out the local historic temples and surfing spots. However, tourism is not an innocent bystander in the South China Sea dispute. Pretending the islands are a tourist destination is a way of normalizing control. Flag-waving tourists are shipped out and, in 2016, China landed two commercial passenger aircraft on Fiery Cross Reef, one from China Southern Airlines and the other from Hainan Airlines.

Despite the bristling weaponry that crowds its concrete surface, a civilian presence is key to convincing the world that Fiery Cross is part of China. In 2011 China Mobile announced that residents of the Spratly Islands (of which there were, in 2011, almost none) would henceforth be enjoying full phone coverage. Fiery Cross is reported to possess a coral-restoration facility in addition to a lighthouse and a hospital. After sinking numerous bore-holes, fresh water was discovered in the vicinity, and in 2019 China's Ministry of Transport opened a maritime rescue centre on the island. The official website 'China Military Online' reports that 'fishermen who go fishing in the South China Sea can also stop on the island to find shelter or replenishment'.

The Chinese occupation of Fiery Cross did initially have non-military aims. In 1988 UNESCO asked the Chinese to build a weather observation station in the region and Fiery Cross was the chosen location. Even this was controversial. Vietnam objected and sent ships with construction materials in a bid to begin their own building work. They were seen off by the Chinese navy, one of the first skirmishes in a fractious cold war that frequently heats up.

Fiery Cross Reef reminds us that artificial islands are not just about leisure, pleasure or offshoring infrastructure; they can be key military assets. Such islands have a long history. Yet their recent history and our ability to now build them bigger, more quickly and further away from the mainland should make us wonder if international law, which gives nations with islands a hugely extended territorial range, needs to be updated. I'd argue that artificial military islands – which are weapons just as surely as the aircraft carriers they resemble – should be excluded from these generous territorial provisions. Otherwise, I fear that many other lonely reefs and shallows will be commandeered and mutilated and the seas dotted with ever more daring and ferocious forward placements.

PHOENIX ISLAND, CHINA

At seven o'clock every night a switch is flicked and Phoenix Island's pod-like towers begin pulsing with multicoloured patterns, swimming fish, exploding fireworks and scrolling celebratory messages in Chinese. Down on the sand of Sanya Bay, still warm after another sweltering day, well-mannered parties are in progress with extended families taking selfies and having picnics.

Phoenix Island has been dubbed the Oriental Dubai, though it's minuscule by comparison and will soon be overshadowed by China's newer artificial leisure and residential islands. It's just off the coast of Sanya on the island of Hainan, 'China's Hawaii', where roadsides are lined with coconut and banana trees. There are ten more artificial islands being built around Hainan's shores, including the stupendous Ocean Flower. They are all leisure and residential islands, absorbing the surplus cash made by wealthy mainland Chinese who fancy a place in the sun. South of Sanya is the South China Sea, a feverish zone of military island-building.

I'm booked into a seafront hostel on Sanya Bay with a fine view over the lozenge shape of Phoenix Island. During the day you can see that, although one half of it is complete and a hive of activity with frequent helicopter traffic, building work on the other half has ground to a halt. Sanya has a laid-back vibe. The honking of buses and scooters is incessant but punctuated by the whoops and giggles of holidaymakers, padding about in flip-flops and clasping umbrellas to keep off the sun. On my first day here, weaving through the beach crowds of a few Russians but mostly mainland Chinese, I make my way over to the bridge that crosses to Phoenix Island. It's a big moment for me. I've been planning this trip for months and spending sickening amounts of money and it all culminates here and now.

There it is: an elegant white road bridge arching across. This should be easy. But closer to it, I see there are swing barriers and half a dozen uniformed guards. I begin to walk by the barriers but I'm waved back. I have been diligent with my introductory Chinese classes but no one understands anything I say. A border official presents me with a laminated A4 sheet that reads 'The island is not open. It is open for guests.'

The next day I book a room online and am waved through. But this minor mishap tells us something. More often than not, artificial islands are not open to the public. As they proliferate so does the idea that public space – places without barriers and guards – is second best. Hainan's new islands point towards a future in which valued places exist as luxury enclaves, well away from the ordinary life of the city.

A golf buggy trundles me and a few other unsmiling residents across the bridge in dead silence past immaculate hedges blazing with azaleas and more guards, now in crisp white uniforms. The final guard, outside Tower D, snaps the driver a stern salute. The hotel is near the bottom of the tower, which shoots up from slender concrete supports. As we cram into the lifts, it is like entering a rocket ship heavy with fleeing people. The island's curving, sinuous buildings are supposed to put you in mind of the sea. The architect who designed them, Ma Yansong, told the *Hainan Daily* he wanted something that looked like it 'grew out of the sea. They should be curvy, just like coral or sea star.' But the scale is way too big for that and the whole thing points skywards. It's a place that wants to escape.

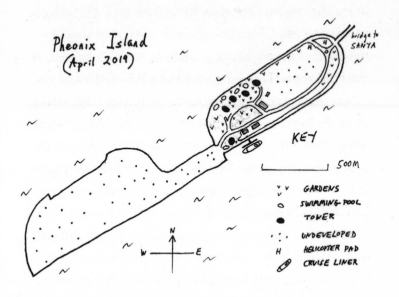

Pheonix Island
(April 2019)

bridge to SANYA

KEY

500M

v v   GARDENS
o   SWIMMING POOL
●   TOWER
. . .   UNDEVELOPED
H   HELICOPTER PAD
⬭   CRUISE LINER

N
W   E

Since I've paid only £40 for my room, I am steeled for a dismal night. Instead, I'm shown into a huge apartment with its own private garden, planted with pretty flowers and inside and outside bathtubs. I flump onto the bed, slinging a friendly arm round a pile of white towels that have been skilfully folded into the shape of an elephant. My eye is caught by a cleaner, burnishing a handrail just outside this princely domain: he's moving along methodically, and when he spots a bit that isn't up to snuff, he applies himself with real energy. I head out, and pass the guard who snapped a salute earlier. He offers a broad smile, gives me a bow of welcome. As so often in China I'm having to rethink. What I took for sternness, even officiousness, was something else: a sense of pride in doing a job well. Like the guy cleaning the handrail. People here aren't just turning up for work; they are part of something: something small and particular but also spectacular and wildly ambitious. One of the most important statistics that explains China is that between 1981 and 2013 China lifted 850 million people out of poverty, with the percentage of people living in extreme poverty (living on under $1.9 a day) falling from 88 per cent to 1.85 per cent. The Chinese are proud of what they have achieved and who can blame them?

The near-disappearance of the worst kinds of poverty was accompanied by a huge increase in the number of people with enough money to afford holidays and second homes. The new wealthy are literally reshaping China, and it is their money that is fuelling Hainan's island-building boom. Pursuing a tourism- and real estate-led development strategy, the provincial government based in the capital of Haikou has been keen to welcome mainland money.

The new islands create lots of opportunities to build high-value, waterfront houses and leisure resorts. Reclaimed between 2002 and 2003, Phoenix Island opened in 2015. One of its key assets is a cruise ship terminal, offering a city-centre berth for the increasing numbers of Chinese who favour a holiday at sea. Before nightfall, I try to find out what else brings people here. There are a succession of very popular swimming pools, long tables decked out ready for an open-air barbecue, and a lot of expensive sports cars revving noisily. It doesn't take me long before I think I've seen everything. The island is set up for families and romantic getaways; the air is warm and full of contented laughter. There is plenty to enjoy here, just not for me. People look puzzled as they catch sight of me – a lone male and the only Western person, it seems, on the island. I try to look purposeful, walking quickly, pretending that I have someone to meet; I stare at the blank screen of my phone, hiding. All this empty performance is made emptier by China's 'Great Firewall', which means that all my apps and servers, from WhatsApp to Google, are blocked. Burning with self-consciousness, I settle down poolside next to splashing families, and tinker with the notes I made in Hainan's capital, Haikou, where I spent last week.

In the sprawling, dusty city of Haikou I stayed in the tallest hotel because I thought it would give me a view of two incomplete artificial islands anchored just offshore. The nearest is called Huludao, meaning 'gourd island', and has bulbous ends and a thin midriff. The low trees that have begun to colonize its scrub reveal that building work has not been continuous. China's new islands are plagued by stop-and-start construction. It's not just a Chinese problem. Expensive, complex projects like island-building are vulnerable: economic conditions change; a contractor goes bust; one level of government gets cold feet and everything is put on hold. The busy circle of boats around Huludao suggests it may be soon back in business. The master plan for the island shows swirling, sail-shaped buildings and a huge central tower that will feature 'ultra-star hotels'.

China's islands often inscribe Chinese culture into the sea. Just along the coast is Nanhai Pearl Island, shaped into a yin-yang symbol. The 'yin' half is to be a residential complex and the 'yang' a marina. On the other side of town is Ruyi Island. (A *ruyi* is a curved ceremonial baton, an ancient motif in Chinese art.) This is the most ambitious of Haikou's new islands: 4.5 kilometres from the shore, covering 716 hectares (more than twice the size of New York's Central Park) and divided up into six entertainment, wharf and residential districts. Finally, there is Millennium Hotel Island. The plans show a small island

The new island of Utopia: illustration from the 1516 first edition of *De optimo rei publicae statu, deque nova insula Utopia* (Of a Republic's Best State and of the New Island Utopia) by Thomas More (*Lebrecht Music & Arts / Alamy Stock Photo*).

Early phase of Chinese building activity on Johnson South Reef (*Kyodo News/Getty Images*).

A number of the Solomon Islands have disappeared or are soon to disappear. This photograph shows the Russell Islands (which lie in the middle of the archipelago), where shorelines are rapidly retreating (*robertharding / Alamy Stock Photo*).

My row boat, unnamed island, Loch Awe.

Snack food packaging, fishing lines and broken glass found on the shore of an unnamed island, Loch Awe.

Multiple cranes slotting together the Kleindienst Group's 'Heart of Europe', as seen from 'Lebanon', 'The World', Dubai.

The islands of 'Palestine', 'Jordan' and 'Saudi Arabia' as seen from 'Lebanon'. In the distance: downtown Dubai with the hazy pinnacle of the Burj Khalifa.

In front of the *Nieuw Land* museum, Flevopolder. A white head rises from a bright blue rectangle representing the sea, with a life-size Cornelis Lely standing on top, punching the air.

Chek Lap Kok, Hong Kong. Walkers' finger post, pointing the way to 'Ancient Kiln' and 'Historical Garden', with Hong Kong-Macau bridge in distance.

Cutter head from a 'cutter-suction dredger', Chek Lap Kok, Hong Kong.

Cable car on Lantau Island, with Chek Lap Kok International Airport in the distance.

Students sitting on the tiled plaza beneath HSBC headquarters, Hong Kong. A security guard has just arrived and is telling Michael that this is not allowed.

Sanya Bay. The inset of the South China Sea as it appears on the Hainan tourist map. The thick dotted line delivers almost the entire sea to China. The squiggles in the centre are the Spratly and Paracel islands.

Phoenix Island's towers pulsing with multicoloured patterns. Sanya, Hainan.

Screenshot of graphic showing the 'Ocean Flower', Hainan, taken from developer's promotional film. Translation: 'Sea Flower Island. Filmed March 2019' (*Evergrande Group*).

Ocean Reef, Panama City. Isla II is on the right connected to Isla I by a bridge. Behind Isla I are the towers of Punta Pacifica.

Gated entry onto the private bridge that leads over to the Ocean Reef islands, Panama City.

with a colossal 'seven star' hotel planted in the middle. When these are all finished Haikou will have a jaw-dropping skyline: big, brash and in the sea. Scott Myklebust, an architect working in Hainan since 2005, told CNN 'The market has been an arms race to develop the next most interesting or extreme project.'

Even the terms for the hotels to be built – 'ultra-star' and 'seven star' – are extreme. What do they mean? Opulence and 'high end' are the mantras of all these islands catering for and, supposedly, creating wealth. No one seems to doubt that there will be enough rich people to fill them. Luxury may be the common denominator but it's getting ever more bizarre. An insatiable expectation of excess has been set in motion, in which ultimately nothing can ever be good enough. Usually I stay in cheap rentals but in Haikou I'm in a Hilton and I'm finding its attempts at luxury a little overwhelming. My room has more than forty lights and light-switches and I can't work out how to turn on any one particular light. The toilet is run on so many sensors that it is on the verge of sentience; I have no control over the lid or the flush because it lifts and flushes when it thinks it is appropriate to do so. I lie in the dark, waiting for the toilet to make a decision.

Escaping my luxury room for the hotel's luxury lobby on the thirty-sixth floor, I've arranged to meet a tourism expert, Dr Fu from Hainan University. He is a young man with a ready laugh, and I like him immediately. I can tell he is not a big fan of Huludao, which is disappearing in the mist below us. 'It creates trash problems on our coast because it is so close,' he says. 'The water quality is very bad and the island makes it worse and stagnant.' He adds: 'The Hainanese are crazy for artificial islands. They want to be like Dubai. Some people say Dubai is very successful.'

Dr Fu explains the bonanza of new islands in starkly economic terms: 'The major industry in Hainan is real estate.' You might think building islands was an expensive way of selling property but, in fact, because the price of urban land is so high, it works out to be a cheap option. It can be ten times more expensive to buy land onshore than to build in the sea. Moreover, in China all city land is owned by the state. This also explains why entrepreneurs are tempted to go offshore. The sea is open to unfettered capitalism in a way other places are not.

Later Dr Fu treats me to a hot-pot lunch. While we are slipping slices of bamboo and jellied duck blood into a simmering cauldron, he talks about the place of islands in Chinese culture, especially in Chinese mythology. Islands were the home of the gods and associated with long life, and were felt to be auspicious. Moving up to date, he gets his smartphone out and shows me clips filmed by a friend who lives opposite the Ocean Flower, which is nearing completion on the west coast of Hainan. It looks like a fully formed high-rise city. From above, the Ocean Flower reveals itself as another Chinese cultural inscription in the sea. It is shaped like a lotus with scrolling leaves, the lotus being a symbol of honesty and purity in Chinese Buddhism.

The opening frame of the Ocean Flower developer's promotional film is a computer graphic but it gives a sense of the scale of the project. The words translate as 'Sea Flower Island. Filmed March 2019'. Canyons of apartment blocks dominate the Ocean Flower's 'leaves' while the lotus itself is a collection of fantasy architectures: there are European-style castles and churches sitting beside grandiose hotels and amusement parks. This is one of the world's most ambitious building projects, about 1.5 times the size of Palm Jumeirah. When complete it will have twenty-eight museums, fifty-eight hotels, seven 'folklore performance squares' and the world's largest conference centre. However, Dr Fu tells me that building work has come to a halt. The central government is unhappy. I'm curious to find out why.

Next day, I make my way over to the university campus to talk to some more experts. You don't have to wander far into Haikou's steaming lines of bumper-to-bumper traffic to see why people might prefer to live on one of the new islands. Aside from the scooters, which come at you from all angles, the driving style is patient but the roads are so packed that accidents abound. Haikou is, indeed, going the way of Dubai: a city of fantastical air-conditioned architecture where no one walks. Ironically, the two professors, Dr Xiong and Dr Li – stylish young women with polished

English – explain that in China Hainan is a byword for tranquillity and calm: 'My shoes are dirty after one day in Beijing but here it's a week.' We are talking in a busy, off-campus juice bar and they are eyeing me with amused concern because Dr Li has given me a lift to our rendezvous on the back of her scooter and it was more fun for her than it was for me. 'You seem afraid of my "*scooter*",' she chuckles, rolling the word like a delicious marble as I mop the sweat from my upper lip.

Trying to seize back my lost dignity, I ask about push-back on all the island-building. They tell me that 'there is no environmental movement here'. This despite the fact they personally would never live on any of the new islands because they are 'dangerous with land subsidence'. The push-back is coming from the top. I hear the same story again: 'The government has stopped those islands.'

Eleven new islands are being built round Hainan. I've mentioned six so far, the remaining five are all leisure islands of similar type, the most unusual being 'Sun Moon Bay'. Sun Moon Bay is already built and consists of one island shaped as the moon and one as the sun. An official news agency promises it will be 'high-end atmosphere grade, similar to the well-known Dubai World Island'. 'High-end' and 'Dubai' – these are the magic words for Hainan's developers. So what's gone wrong? The central government in Beijing has, it seems, been raining on the developers' parade and in 2018 it started cracking down on private reclamation projects all over China, citing a lack of environmental assessments and 'proper permits'. The new islands were said to be creating coastal erosion, silting up rivers and damaging ecosystems. According to the Xinhua News Agency and ChinaDaily.com, developers have been ordered to 'restore the environmental damage they have caused' and 'fulfil environmental restoration work as soon as possible'. Nearly all the island projects around Hainan have been officially suspended and 'ordered to carry out environmental impact assessments'. It's not just the developers who are in the dock. In fact, the main target may be local politicians: 'government officials who were found to have violated the law in those cases will be punished'.

I suspect this story isn't just about Beijing's new-found enthusiasm for environmental protection. It's also about cutting developers and provincial politicians down to size. Reining in confident and wilful regions is something the central government does a lot. Yet there is a momentum to Hainan's new islands that won't be halted. It's too late for that. The official slapping-down from Beijing is a temporary set-back, nothing more. The party on Phoenix Island is already in full swing and Hainan's other islands are not far from completion. China is building big – on a scale never seen before.

Back home in Newcastle, I tell people where I've been: Hainan, Sanya, Haikou. 'Where?' No one has heard of them. They will soon.

OCEAN REEF, PANAMA

Ocean Reef is a pair of Dutch-designed artificial islands that jut out from Punta Pacifica, a high-rent and high-rise neighbourhood in Panama City. It is an ultra-secure retreat for Panama's wealthiest families. Joined to the mainland by a permanently guarded causeway, Ocean Reef is not just gated, it is locked down: it has its own marine security force and is surrounded by underwater sensors that pick up anything or anybody over 40 kilos. 'Yes, it is like James Bond,' chuckles James, the affable Scottish-Nicaraguan real estate agent who is showing me round. (It turns out that James's father came to Central America from Glasgow to mine for gold.) The army of service workers that keep the place ticking over have a dedicated tunnel discreetly positioned at the entrance of 'Isla I'. Once below ground, they swap over to electric vehicles, keeping the islands quiet and free of car fumes. Anyone who has been to Panama City will know just how precious that is.

This is my first time here. The sleepless flight from Amsterdam banks down over the Panama Canal and then swings in front of finger-thin skyscrapers that form a crenulated silver wall stacked up against the blue bay. Ocean Reef is already visible, a pincer shape snapping out from the vertiginous glass and concrete shore. Slogans from its website are playing in my head: 'Live an island lifestyle within the city'; 'As we evolve, our homes should too'; 'The first man made urban islands in Latin America'. Even from this height you can tell it's something special. You'd have to be well off to live in one of those extravagant shoreline towers. But there's rich and there's *rich*. Ocean Reef – low-rise, separate, secure – caters for the latter.

I'm arriving in late October 2018 and, though the islands are built, many of the surface features have yet to be put in place. Soon the open water between the two islands will cradle a 200-slip marina designed to accommodate vessels up to 90 metres (300 feet) in length. At that size so-called 'yachts' are more like small ships. But it's not the size of the boats, nor the multi-million-dollar price tag of the apartments, that informs you this is a place apart. Far more telling is the fact that Panama's constitution, which forbids marine estate being sold for private use, was controversially waived on a one-off basis to allow Ocean Reef to be built. Panama's elite have had this place constructed for themselves and they have made sure that no rival islands are ever going to spring up next door to spoil the view.

Panama is the bridge of the Americas and provides a stark divide: on the Pacific side artificial islands are being magicked up for millionaires while, 70 kilometres to the east, subsistence farmers are watching their ancestral islands disappear beneath the waves.

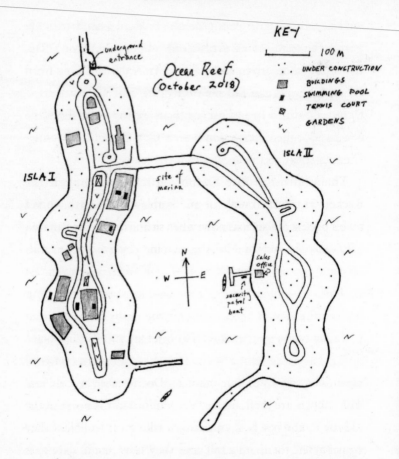

A little smaller than Scotland and just as hilly, Panama is a raw and intoxicating country. Per capita, it's the richest nation in Latin America – far wealthier than any of its neighbours. Yet outside a few pockets, it looks poor. Its capital is claimed to be one of the most ritzy and cosmopolitan cities on the continent, but it's not far from thick jungle and most of the city remains off-limits to tourists. Once you get off the Pan-American highway, driving anywhere in Panama is a trial of potholes and scrub roads. I have, however, been disabused of one popular myth: the Pan-American highway comes to an abrupt halt before Panama's legendary Darién Gap *not* because of impenetrable terrain or wild tribes but due to a long-standing feud with Colombia, the country Panama separated from in 1903.

Thankfully, for once I'm not on my own. My partner Rachel has travelled with me and is able to hold forth while I take a back seat – such as now, when sitting in a bar in Panama City's colonial old town, as we are joined by a talkative local. Dorset-born and covered with tattoos of English punk bands, he works for the minister of tourism and is delighted to find some fellow Brits to while away a couple of hours. He and Rachel are having a great time and, jet-lagged and dozy, all I have to do is offer an occasional encouraging smile. I do perk up on hearing a familiar mantra: 'Don't leave the cobbles.' The cobbles are the road surface you find in the heart of the old town, and it seems tourists are asking for trouble if they stray beyond them. It's a tall order as 99 per cent of the city is beyond the cobbles, but it's also a serious bit of advice. A few hours earlier, we were wandering 4 or 5 metres off the main drag when a young woman in a white pick-up pulled over and leaned out to give us the same warning: 'It's not always safe in those alleys,' she smiled.

Fears about safety guide our footsteps and shape the city. If we had wandered a bit further we'd be in El Chorrillo, a ramshackle, densely packed neighbourhood that was once home to General Manuel Noriega, the drug baron and de facto ruler of Panama in the 1980s. The USA invaded Panama in December 1989 in order to oust him. In the first thirteen hours of the invasion US planes dropped 422 bombs, a lot of them on El Chorrillo. Casualties are still not known but the figure is usually put at about 7000.

The invasion is still a sore point in Panama but, as in many small countries, hostility to the regional superpower is a luxury few can afford and, in the end, security is what counts. We all try to live in islands of safety. Some of us can turn that metaphor into a reality. Drive across the Ocean Reef causeway and the tension, the nagging anxiety, of living in a dangerous city melts away.

The construction of Ocean Reef was thorough. The seabed was scraped down to the bedrock and barges full of rock from the mainland were piled up in the water, then sprayed with sand. There were going to be three islands, each named after one of Columbus's ships: *Niña*, *Pinta* and *Santa Maria*. But in the end the Dutch designers recommended that two would create a more durable shape. Today they go by the prosaic names of Isla I and Isla II. Each is an organic bulging loop, with the second and further island curling just enough around the first to create a protected anchorage.

The residential plots were quickly sold and the first foundation stone was formally laid in 2010 with much fanfare. *La Estrella de Panamá* newspaper reported that 'It was a gala where the cream of society met together with important figures of the current administration' including President Ricardo Martinelli. 'In the front row were entrepreneurs,' *La Estrella* continued, 'accompanied by the Minister of Economy, Alberto Vallarino, who among light music, caviar, wines and champagne celebrated the start of the project dubbed the "Dubai of the Americas".'

The words 'exclusivity' and 'luxury' are front and centre in Ocean Reef's promotional bumf, often followed by 'tranquil' and 'private':

Ocean Reef residents are only those who enjoy the finer things in life, for them quality is not a desire, it is a reality. With this person in mind, Ocean Reef Islands have been created. Being a member of the Ocean Reef community is a privilege. Ocean Reef will be the only place in the city where you can engage in a true island experience – tranquil, private.

The Ocean Reef website rhapsodizes in English and Spanish that 'Exclusivity and Luxury have never been so obvious in any development' while also reminding likely buyers that this is not going to be a mere resort, populated by renters and blow-ins; the properties are freehold and the buyers are nearly all Panamanian. The implication is that they will be passed down through the generations. I was told that few, if any, properties will come back on the market again.

The promotional spiel goes on to talk about the islands as an 'unspoiled refuge' set in the 'midst of timeless natural splendor'. The idea that artificial islands give access to 'nature' is a paradox I keep bumping into. When you arrive on urban exclaves, like Ocean Reef, it's easier to understand. After negotiating the rubbish-filled streets and the built-up canyons on the mainland, on Ocean Reef you can at last see the sky and stare out to sea. You can also walk on the ground. Contact with the earth is an unregarded part of our relationship with nature, but in high-rise, traffic-clogged cities it is an increasingly rare thing. To be able to walk out to see a neighbour – not in some air-conditioned tube but simply round the corner – is the kind of simple pleasure that islands like Ocean Reef allow and put a high price on.

There is another and even more rarely voiced kind of exclusivity at work in Ocean Reef, which has to do with Panama's relationship to race. Panamanians like to claim that theirs is an inclusive, colour-blind country. It isn't. The Ocean Reef publicity shots show exclusively young and good-looking white people relishing its pleasures. They mirror a wider prejudice: every billboard, every TV advert displays nothing but white people. That's striking because Panama is a very diverse country where nearly three-quarters of people are mestizo (mixed). Twelve per cent are indigenous and there is also a large black population, many of whom descend from Caribbean workers brought over to dig the canal. As a rule of thumb, the richer the neighbourhood the whiter-looking the inhabitants. Some words of caution, however: this doesn't mean that Ocean Reef is just for whites; it isn't. Or that 'gringos' – a term applied to all foreign whites these days – are part of the club; they aren't. Panama's racial dynamics are complicated but also stark. The flight to live in secure compounds like Ocean Reef has racial baggage: it's a flight away towards the safety of something not exactly white, but certainly whiter.

Visiting Ocean Reef is not straightforward. You can't just wander up. I know that for a fact, as I tried it. I walked under the entrance arch – adorned with the slogan 'A lifestyle change begins with a vision and a single step' – onto the pristine causeway lined with palm trees. Adopting the brisk 'I know where I'm going' stride that has ended in humiliation many times before, I strode past the guarded booth and the barrier arm. Perhaps I was trying to deploy some white privilege. If so the guard wasn't impressed. He stepped out and called me back, finger-wagging and repeating '*Imposible, imposible, imposible.*'

Thankfully, I have a Plan B: an appointment with James the real estate agent in his offices in Punta Pacifica. I really didn't know how I'd ever get my foot in the door of Ocean Reef and so I emailed him under the pretence that I was 'interested' in buying an apartment. It is a lie that starts to come under pressure even before I arrive. Rachel has agreed to come with me and is bringing with her an ethical agenda I was hoping to avoid: 'I'm not pretending to be buying anything; that would be ridiculous.' The atmosphere in the taxi on our way across town was tense.

We are ushered into a plush back room. Everything is clad in black leather, photos of the local 'Trump Tower' decorate the walls, and beads of sweat are pricking my forehead. I run through a few scenarios – 'We'll need to know what you can afford, señor' – and a lot of them don't end well. A well-built, smartly suited and roughly handsome man joins us, clearly in charge. James is smiling warmly and my butterflies take a rest. He knows – he always knew – that I was not a client. I guess he is intrigued and keen to show off Ocean Reef.

We chat about the 'sinking' islands on the Atlantic side of the country and he readily admits that people in Panama are not that interested in sea-level rise or in climate change more generally. 'Perhaps they should be,' he muses. In this part of the world, the environmental disaster that matters is earthquakes. James opens out the colourful master-plan document of Ocean Reef and explains how computer modellers subjected the islands to 14,000 earthquake simulations. It is built to withstand anything the Ring of Fire can throw at them, which also helps explain why, rising to over 9 metres, Ocean Reef has the highest ground on the seafront.

Faced with James's friendliness, I am disarmed; more than that I'm eager, laughing and agreeing with everything he says. Rachel casts me a quizzical look; 'I was wondering how low you could go,' she tells me later. We climb into James's superior four-wheel drive, the barrier swings high, and our stately progress is met with a series of greetings and nods from roadside workers. Once on foot, the waves and greetings continue. They are all from receptionists and service workers since only twelve families have so far moved across. Isla I is not finished and Isla II (apart from the main sales building) is bare land. We take a lift onto a series of connected roof terraces and I become aware of piped music drifting through the warm air. We wander past infinity pools. I make appreciative noises, though I'm more interested in the odder sights, such as the walls plastered with plastic green foliage and a rooftop 'pitch and putt'.

The sky is overcast and threatening rain and I realize I'd better take a few photographs before it starts pouring. (It rains a lot in Panama: two and a half times more than in Britain.) I snap one of Isla II, mostly undeveloped and connected to Isla I by the bridge that will form the apex of the marina. To its left the high towers of Punta Pacifica rise in the distance, including the distinctive arc of the JW Marriott hotel (formerly 'Trump International Hotel and Tower'), foregrounded by the palms and low-density, low-rise homes of Isla I.

Back in the car, as we roll past a well-equipped and pristine children's playground, the conversation turns to the island's family ethos. James will bring his own young family to live here when the island is complete, and he has no doubt that Ocean Reef will function as a real community. It seems likely, since a lot of the families – there will be 400 in total – who will come here already know each other. It's 'selling to friends', says James.

To keep the community cogs turning, a $1000 monthly service fee buys residents into the 'island app', which is used to book gym or restaurant time and access other facilities. We drive over the bridge that leads to Isla II and park up by the sales centre. I've become nervous again: do I have to pretend to be 'interested'? However friendly and forgiving James might be, I feel like a fraud. If I sold everything I own I couldn't afford half the price of the cheapest apartment here. Yet here I am, swanning around, pretending I'm used to this sort of thing. I amuse myself with the melancholy thought that this sums up a lot of what I do. It's certainly par for the course when visiting artificial islands. As Rachel and James chat away, I try to make myself inconspicuous, taking a studious interest in various glass-boxed models that exhibit the master plan, the marina and different apartment blocks where tiny people dash about in bright casuals.

One of those tiny people could be me – waving to friends, popular at the waterfront ... My reverie is broken: James has one final sight to show us. We walk out to a pontoon where Ocean Reef's security patrol boat is moored and take in the spectacular view. Punta Pacifica rises vertically to one side and on the other are the dark humps of the Pearl Islands, an archipelago that lies 50 kilometres or so out in Panama Bay. It is incredible: utterly urban yet so far away from the stress of the honking metropolis.

Next day, back in the city, a yellow taxi pulls up and we begin negotiating, trying to get somewhere we want for a price that seems fair. It's the usual thing with cabs in Panama City and it's an unpredictable process. This morning's journey is a success. It takes us round El Chorrillo, past street corners where men are making sugar-cane drink from ancient crushing machines, and past parties of Kuna women recognizable in their bright shawls and leg bangles. Our destination is a hill that overlooks both canal and city, on the top of which flaps the world's largest Panamanian flag. At one point a man calls us back and jabs his finger up: we have just walked under a branch on which a fat boa constrictor is looped, sleeping off its last meal.

At the crest of the hill a huge panorama opens out. On one side rows and columns of colourful cargo containers stand waiting for loading by the Panama Canal; on the other is the drama of Panama City. And there's Ocean Reef, reaching out into the sea – part of the city and yet free from the city, in it but not of it.

Natural, Overlooked
and Accidental:
Other New Islands

Natural

HUMANS BUILD VERY small, flat islands. The planet is engaged in larger construction projects: such as colliding the 103 million square kilometres of the Pacific plate into the 47 million square kilometres of the Australian plate. The former is being shoved under the latter, where it heats and melts, lava bursting to the surface to create mountainous islands. The most spectacular natural island to emerge in recent years was one of them: Hunga-Tonga.

As we worry and wonder at what people are doing to the planet, it's easy to forget that, in geological terms, we are not that big a deal. Imagine you are trekking towards a great mountain. It's looming ahead, a dark immensity. As you begin climbing you stare down in horror to find your boots are bustling with ants. Gazing about, you realize the whole mountain is covered with their trails: everywhere they are shifting and shovelling, creating huge farms and eating away at the vegetation, amassing anthills dark with tens of millions of insects. It seems odd to anyone gazing on the great mountain from any distance, but the ants think of themselves as the centre of creation: that this mountain – everything – is all about them. But being clever creatures, they have collected data and know for sure that it is, indeed, the 'era of the ant'.

Today geologists talk about a new geological era defined by human impact on the planet – the Anthropocene – though they cannot agree when it started (some say with industrialization; others believe with the first dusting of nuclear radioactivity). It's plausible: human activity *has* transformed the planet's climate and landscapes. According to Owen Gaffney of the Stockholm Resilience Centre, 'we move more sediment and rock annually than all natural process such as erosion and rivers combined'. But defining the planet around us may not be so bright. It implies that if we screw things up we can click our fingers, do something clever, and fix it again. It is necessary to be reminded, time and again, that we are dependants on rather than masters of Earth and that, even as we harm it, the planet keeps on turning and will continue to do so well after the last of us is not even a memory. And whatever we do, the Earth will continue to make islands.

Natural new islands are produced in two main ways: through volcanic activity and through changes in sea and land levels. The latter may create the most islands but the drama, beauty and headlong pace of the former is intoxicating.

The Earth is divided up, to a depth of between 15 and 200 kilometres, by seven major plates and an awkward bunch of microplates. The continents and oceans sit on these plates. They are a shifting, buckling jigsaw and they are very hot. Recent research has shown that the Earth's core is about the same temperature as the Sun. Heat energy melts the outer core (the inner core is now thought to be solid) and drives the motion of the plates. Islands are formed both where plates crash together and where they are ripping away from each other.

The Atlantic is being zipped open by a mid-ocean rift and has about twenty active volcanoes, all of which are building undersea mountains (called seamounts). The largest island created so far by this process is Iceland. The highly active nature of volcanism in and around Iceland – most famously the creation of Surtsey in 1963 – suggests something else may be going on. It turns out that Iceland is not just on a volcanic rift but sits above a 'hot spot'. Sometimes called 'anomalous volcanism', hot spots can occur anywhere and are often found well away from plate boundaries. Why magma spews up at them is still not known. The most famous hot spot is under Hawaii (other examples are Cape Verde and Galapagos), a highly volcanic island chain that is thousands of kilometres from the nearest plate boundary. In 2018 the Kīlauea volcano on Hawaii's Big Island, which has been erupting almost continuously for over three decades, flung up a new islet. It caught people by surprise. Such islands always do. We don't know the when, where or entirely why of hot-spot volcanism. What we do know is that hot spots are not fixed in place, that they rove around, and that they are not single sites but extensive. Like most volcanic islands, Hawaii is the visible part of a long chain of undersea volcanic seamounts. The 'Hawaii chain' is 5800 kilometres long and includes hundreds of undersea mountains. Some are rising but others are eroding, submerged islands that came and went long before humans were about to tell the tale.

Swim west of Hawaii, and keep going for some 5000 kilometres, and you will arrive at one of the most active zones of island production: the crunch point between the Pacific and Australian plates. It is a complicated place: in between the big plates a number of geological oddities – microplates – shove and butt up against each other. The Tonga microplate is the world's fastest, moving at speeds of up to 24 centimetres per year. Here we find the newly risen Hunga Tonga. It's not clear how long it will last; volcanic islands come and go. Another Tongan ephemeral island is Home Reef, which came into existence and disappeared again soon after volcanic eruptions in 1852, 1857, 1984 and 2006.

There are many volcanoes bubbling under Tongan waters as well as the world's second-deepest canyon. Volcanic islands often form alongside long trenches, creating the classic island arcs seen in the Pacific and Caribbean. The deepest point of the Tonga Trench, 'Horizon Deep' is 10,800 metres below sea level (for comparison, Mount Everest is 8848 metres high). Horizon Deep is only 100 metres higher than the world's most famous plate boundary abyss: the Mariana Trench's 'Challenger Deep', formed in the north Pacific and the deepest place on the planet. The best known of the volcanic islands near the Mariana Trench is Nishinoshima, which – after eruptions in 1974, 2013 and 2017 – is now nearly 3 square kilometres in size.

The most surprising form of 'volcanic island' is the floating one. One of the types of rock produced by undersea lava streams is pumice. It's so light that it floats, creating 'pumice rafts'. The largest pumice raft ever recorded was spotted in the South Pacific by the Royal New Zealand Air Force in 2012, 1000 kilometres off the Auckland coast, spread over an area of 25,900 square kilometres – or 'nearly the size of Belgium', as the New Zealand press described it. Pumice-raft islands soon drift apart but it has been speculated that animals and plants may sometimes hitch a lift and hence colonize new shores.

Island-building is often a combined effort. In the warm waters of the tropics many 'volcanic islands' are actually old, worn-down volcanoes built up by coral, an industrious animal that also creates offshore reefs. Coral can accumulate to considerable depths, being well over a kilometre thick on some Pacific islands. Another reason why 'coral islands' appear is that the weight of nearby volcanoes pushes them up. When you stomp down on a surface, you'll see the area around the impact rise up. It is the same with the Earth. Such pushed-up reefs are called 'makatea islands', named after the Tahitian island of Makatea, an ancient reef that the weight of nearby rising volcanoes has caused to lift up out of the water.

Yet most new islands are created not by volcanic activity but by changes in the level of the land and the sea or by the way waves and wind shift shorelines. Islands created in this way are so common and usually so short-lived that they are rarely named or inhabited. Most are the result of deposition: sediment rushed along rivers often creates new islands in slow-moving stretches of water or out to sea. Others can emerge over just a few days: sand-bar islands shunted into place by waves and storms. Inland, long periods of drought may give birth to another species: drying lakes usually reveal a mottled flock of brown, dowdy islands.

Sea levels have risen and fallen for millions of years, creating and destroying millions of islands. If sea level continues to rise at the rate it is currently projected, then we will see the break-up of coastal areas, such as the eastern states of the USA, as well as the fragmentation of islands created by ancient sea-level rise, such as Britain, into archipelagos.

We are currently in an 'interglacial' or warm period of the Earth's climate, a period that started some 11,700 years ago. The ice retreat and the flooding that we are seeing today compounds an existing, natural process with an unnatural, modern one. Artificially accelerated global warming is already producing many new islands, though they rarely make the headlines. On Russia's long north coast, a group of nine islands in the Novaya Zemlya and Franz Josef Land archipelagos were added to the map in 2015 by the Military Topographic Directorate; almost every year more are added. They have emerged thanks to retreating glaciers and melting sheet ice. The biggest of the new islands is 2 kilometres long and 600 kilometres wide. It is predicted that ice retreat will soon reveal that Spitsbergen, the largest island in the Norwegian archipelago of Svalbard, is not one but two islands, with open water separating the island from what was thought to be the peninsula of Sørkappland. Already, in the west, ice melt has revealed that Spitsbergen's 'Flower beach peninsula' – or Blomstrandhalvøya – is an island; it has been renamed Blomstrandhalvøya Island.

Islands created by melting ice are appearing very fast. Those caused by the land rising are much slower affairs. Around twenty thousand years ago ice sheets covered much of northern Europe and North America. The weight of all that ice pushed the Earth's crust down by up to half a kilometre. Now that much of this ice is gone, the Earth is readjusting itself; it is bouncing back. At the present rate, in two thousand years or so the Gulf of Bothnia, which separates Finland and Sweden, will close up in the middle, turning its northern arm into a lake. The Kvarken Archipelago, which lies halfway down the Gulf, is the most spectacular example of 'rebounding' land. It is made up of 6550 low islands and counting. After new land first emerges, it takes about fifty years for it to grow large enough and to dry out enough to become usable for house-building.

Although uplift creates islands, in the long run it leads to their disappearance. As the water drains away, archipelagos are turned into hilly landscapes. Off the coast of Juneau – in Alaska, and today connected to it by a long bridge – is Douglas Island. This island is steadily joining the mainland; the channel between it and Juneau is silting up. One day Douglas Island will be an island no more. When that day will be is getting harder to judge because of the way global warming and rising sea levels have complicated things. At the moment it is predicted that the phenomenon of new land creation in the far north will continue but at a slower rate.

Overlooked

In 2015 it was revealed that Estonia had 2355 islands rather than 1521, as previously thought. In 2016 the Philippine national mapping agency revised the total of islands that make up the archipelago, adding 534 to the previous tally of 7107. These aren't literally 'new islands'. In part, what is happening is that satellites and aerial photography are showing a much fuller and more detailed picture than we have ever had so that overlooked islands are being added to the map. It's also true that national mapping agencies the world over are getting keener to identify and claim as many islands as possible.

Estonia has lots of rocky islands in the Baltic Sea, the great majority of which are uninhabited. Some of its new islands are likely to be the result of glacial rebound, but most are the product of politics not nature. A small country in the far north-west of what was formerly the USSR, Estonia was once a very minor part of a huge empire and Soviet cartographers did not make detailing its shores their life's work. With Estonia's independence in 1991, a more scrupulous approach began to be taken to the national map. Since countries can claim a 200 nautical miles 'exclusive economic zone' around their islands, the incentive to bag new ones is considerable. That the new islands are sources of national pride was driven home by Estonian Public Broadcasting, whose report on the story concluded with the sardonic observation that it will 'come as a further blow to neighbouring Latvia, which has a famously low number of islands, officially at one, and that too is man-made'.

Only 318 of Estonia's islands are larger than 10,000 square metres, which begs the question: how small can an island be before we stop counting it as an island? The Filipino cartographers adopted the sensible idea that an island has two essential features: some part of it must be above high tide, and it must be able to support either plant or animal life. They need to visit their new discoveries in order to provide 'ground validation' of these two points. But does every stone jutting up from the sea at high tide count as a separate island? The requirement that an island sustains life points towards Article 121 of the United Nations Convention on the Law of the Sea. Here it is stipulated that 'rocks which cannot sustain human habitation or economic life of their own' might be islands, but 'shall have no exclusive economic zone'. Yet few islands today 'sustain human habitation or economic life' on their own, while almost every rock can be found to support some kind of life.

If you ask how many islands are in the British Isles (the main islands of which are Britain and Ireland), the number you will get will vary enormously. One recent definition came from a retired marine surveyor called Brian Adams, who suggested that an island is at least half an acre. If this is the case, then there are 4400 (210 of which, Adams says, are inhabited). However, if you go to Wikipedia you'll find that there are 'over six thousand' and that 136 are inhabited. The wrangle over how many islands are in the Minquiers and Écréhous reefs (which lie, respectively, south and north of the English Channel island of Jersey) is a lesson in how tiresome island-counting can become. These reefs have long been claimed by France but they have a huge daily tidal range and what is permanently above water varies from season to season, year to year. It took thirteen years of talks between France and Britain to arrive at an agreed demarcation. One Jersey politician involved in the negotiations described it as literally counting the Minquiers and Écréhous 'rock by rock'. The border agreement came into force on 1 January 2004. It was only then that the boundary between France and Britain was finally fixed.

The hope that all our planet's islands can be counted finally and irrevocably falls apart in Russia and Canada. You can get a sense as to why from the name of the eastern shore of Georgian Bay, which is just one arm of Lake Huron: Thirty Thousand Islands. That is a rough estimate of the number of islands that dot this low-lying, pine-clad part of Ontario. It is the world's largest freshwater archipelago, but it is also a big clue that trying to count all the world's islands may be a fool's errand.

Accidental

Many of the islands created by human activity were neither designed nor foreseen. They occur as an accidental consequence of quarrying, mining, dredging, reservoir-building or chucking waste into the seas. Pebble Lake in Hungary, where tiny islets crowded with holiday homes are surrounded by the cold waters of a flooded quarry, and the Trash Isles, are examples of accidental islands.

The fugitive island of New Moore helps us think about what 'accidental' means. New Moore emerged a few kilometres out to sea in the Bay of Bengal after Cyclone Bhola in 1970. Created on the border between Bangladesh and India and coveted by both, it looked for a while like New Moore would be a battleground. The Indians stationed troops on the island in 1981 and ran up the Indian flag. The conflict never came, however, largely because New Moore began to disappear; by March 2010 it was fully submerged. Many interpreted this story as a faintly absurd example of people squabbling over something that nature had created then destroyed.

River sediment creates islands, and natural land subsidence is gradually lowering the land under and around the Bay of Bengal. But sea-level rise and river sediment is also being affected by human activity. Road-building upstream contributed to New Moore's creation by triggering landslides that added sediment to the river. Deforestation across the region, especially the felling of mangrove trees, is also changing the coastline, denuding them of protection and stability and making it more likely that islands are created far out to sea.

New Moore was an accidental island but it was also a product of nature. The story of its short life shows us that, for many of the new islands that suddenly appear on the shores of our crowded continents, untangling what is natural and unnatural has become impossible.

HUNGA TONGA-HUNGA HA'APAI, TONGA

Over the last 150 years there have been only three volcanic islands of any size that have sprung from the sea
and survived more than a few months. One is Anak Krakatau
('Child of Krakatoa'), which had been growing since 1927
between Sumatra and Java until, in December 2018, an eruption caused two-thirds of the island to shear off into the
sea. The second is Surtsey, south of Iceland. The third, since
December 2014, is Hunga Tonga-Hunga Ha'apai.

In their early phase, volcanic islands don't just rise,
they *writhe*; shorelines and hillsides spasm and flex with
every passing week. There is an awful struggle to be born
that is accompanied by shrieks and subterranean booms.
Hunga Tonga-Hunga Ha'apai lies on the western fringe of
the Pacific archipelago of Tonga, an ancient kingdom that
comprises 169 small islands and sits just west of the international dateline. Tonga straddles a western arm of the Pacific
Ring of Fire, one of the world's most volcanically lively and
earthquake-prone tectonic collision zones.

On 19 December 2014 an undersea eruption began 45
kilometres north of the Tongan capital, Nuku'alofa. Early
in the new year of 2015 a new volcanic island was sighted.
Today the island is about 3 kilometres long and looks something like a fat bat, with two rocky wings and a pendulous
belly in which reposes a round crater lake. The wings of the

bat are two pre-existing, remote, uninhabited and (for the Tongans) famously earthquake-prone islands: Hunga Tonga to the east and Hunga Ha'apai to the west. The first signs of the new island formed in the sea between these two protecting arms. Eventually it grew so large that it sprawled over and connected them up. It still has no official name, though the scientific community has taken to labelling it Hunga Tonga-Hunga Ha'apai.

I'm cocooned up in a beachside bungalow in Nuku'alofa with the first person to land on Hunga Tonga-Hunga Ha'apai, the fabulously tattooed and sinewy Pacific mariner and rum distiller, Branko Sugar. Outside, a warm wind is making the long hard leaves of the coconut and palm trees clack: Cyclone Keni is on its way, arriving just a few months after Cyclone Gita tore roofs off across the country.

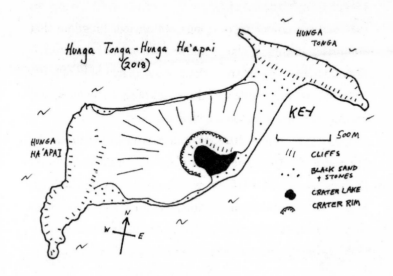

I've been told there will be a short lull, a window of maybe twenty-four hours or so, in which it will be safe to sail to the world's newest volcanic island. That's tomorrow. For the time being I'm happy to be held captive by Branko's stories about what we can expect when we arrive on what he calls 'Hunga Tonga' (this is the name Hunga Tonga-Hunga Ha'apai goes by in Tonga so, from now on, I'll use it too).

The way Branko tells it, Hunga Tonga is a prismatic and strange place and not as terrifying as I'd feared. His voice carries many journeys: a strongly flavoured, fifty-eight-year-old, Croatian-Italian-Swedish-Tongan brew. 'We were the first to come there,' he tells me. 'When we came back I got phone calls from New York, ABC, NBC and god knows what. I freak out and just hung up the phone.'

Branko begins to explain, not for the first time, that to get onto Hunga Tonga we'll have to moor 10 metres offshore and swim the rest of the way. He turns a wary, weather-wrinkled eye my way and is clearly wondering if I'm even capable of getting off the sofa. Unsure how to prove my athleticism, I proudly show off the little Tupperware tub that, for weeks now, I have practised cramming with my essentials (phone, camera, water bottle) and will loop round my waist: just so!

'Yeah, sure. Fine, fine.' This is not a man who does Tupperware. It's the island that matters. 'Ah, that lake! Where ground zero was!' Branko makes a cannon noise and continues: 'It's a green colour; it smells a bit green like green paint, and the ocean is deep blue so you can stand' – now he leaps up, filling my small room – 'Yes! With the deep-blue ocean on this side and the green, green lake on the other.' It's glorious. I'm a wide-eyed child as the island is conjured before me. 'Plants? Yes, yes, plants is already growing. We planted six coconuts and I saw them last week and they are growing. But normal trees, grass have started growing. There is thousands of birds, eggs, little chicks. Everywhere, everywhere on the ground. You can see it from the boat, it's green.'

We scroll through some of Branko's photographs. Many of them show the ground turning to thin scrub and littered with simple nests and a big blue sky, full of gulls.

In brooding retrospect, I realize that the fierce glamour of new volcanic islands has been beckoning me for a long time. I was born a year after Surtsey, the most filmed and famous one of the twentieth century, broke through the North Atlantic in 1963. Named after Surtr, a fire demon from Norse mythology, Surtsey was a visual sensation. Unforgettable films showed burning red entrails of lava. Some of these molten streams were not red but black and crusted and torn with terrible golden wounds. They spewed quickly seawards, exploding into white steam. It was horrible and beautiful, elementally powerful and disorientating. To *see* that I shared a planet with this otherworldly entity made the suburban streets around me feel fragile and unreal. The taming of Surtsey by plants and animals had its own primitive allure, like bringing first life to a dead planet. The island's first plant, a sea rocket, sprouted in 1965; its first bird's-nest was found in 1970. Today, Surtsey is a plump, naan-bread-shaped island with swathes of green scrub-land and thriving bird and seal colonies. Surtsey is protected from humans. No visitors are allowed, other than authorized scientists. Used as a laboratory for the study of natural colonization, the island is sealed from contamination. A website devoted to its study primly reports that 'It is believed that some young boys tried to introduce potatoes, which were promptly dug up once discovered.' Still more shocking: 'An improperly handled human defecation resulted in a tomato plant taking root which was also destroyed.'

There is no chance of such consideration being extended to Hunga Tonga. Human defecation – or Branko's coconut-planting activities – are the least of its worries. Tonga sits on a trans-Pacific drug route, and I've been told that packages of cocaine and other narcotics regularly litter the beaches of its numerous remote islands.

Hunga Tonga was visited by NASA scientists six months after my trip with Branko. The story got quite a bit of press coverage. I couldn't help noticing the huge size of the yacht that took them there. Since the island's first appearance, the international geological community has been using satellite imagery to trace its growth. The evolution of Hunga Tonga is even providing a rare glimpse into volcanism on other planets. For example, it has been used to model 'small kilometre-scale hydro-volcanic edifices in the north plains of Mars', to quote one scientific paper lead-authored by NASA researchers.

Hunga Tonga sits on the rim of an underwater volcano that rises 1400 metres from the sea floor. The last major eruption here was in 2009, an event that also saw lava shoot up to the surface, adding a new shore to the island of Hunga Ha'apai. Since it is made mostly of ash, early predictions were that the new island of Hunga Tonga would soon be washed away. This hasn't happened. Although it is being eroded at about five times the rate of Surtsey, some of its mounds of ash appear to have mineralized and hardened. Hunga Tonga is far more robust than once thought and its predicted life is now a guessing game, with estimates of anywhere from seven to forty-two years.

Into the new year of 2015 Hunga Tonga was still vigorously pouring out clouds of fine debris, a spectacle easily visible from Tonga's capital. The Tongan newspaper *Matangi* reported a series of natural wonders: the sea was turned 'frothy white, chocolate and red while the sun shone through a champagne sky'. 'Pretty bizarre out here this morning,' commented one beachside resident to a *Matangi* reporter: it 'started out ordinary enough then the beach misted over with a brown haze and the ash cloud created rings around the sun'. That the sea turned red was reported by multiple sources and remains an enigma, though it is widely thought to have been caused by algae responding to high water temperatures.

The first vessel to report a sighting of the new island was a ship of the Tongan Royal Naval, on 14 January 2015. The captain also noted that the volcano, now above the surface, was erupting every five minutes. The first few months of Hunga Tonga's infant life saw it morph in size and shape. After a period of rapid expansion, when it latched on to its western neighbour, it began to shrink. In May 2015 waves washed away the bar of material separating the island's crater from the ocean. However, this debris did not all disappear off into the ocean; a lot of it was carried eastward, joining the island to its eastern neighbour. Branko saw the transition unfold. 'When the eruption died out and we went there,' he told me, 'the new island was connected to just one other island, so there was still a little water between it and the other one. The next time we came, four weeks later, that little channel was now land and it's still like that now.'

Photographs taken on Hunga Tonga in 2015 show an ash-grey and black lunar landscape. Anything living on the two rocky islands it swallowed up had been burned away. The landscape was colourless, with sweeping charcoal hills caked with lines of ragged gullies. Walking on the broken surface was very difficult and some experts warned against venturing onto the island at all in case the crust collapsed.

The world's news media has an unslakable thirst for extraordinary images that keep as many internet surfers as possible clicking through. Dramatic volcanoes come with a guaranteed income stream. Even small, ephemeral volcanic islands are shoved onto the geological catwalk. In 2018 one a mere 8 or so metres across, just off the coast of Hawaii, was being splashed as news. A journalist from the *Washington Post* emailed me with a list of questions, such as: 'Are we likely to see more and more new islands forming?' With the profusion of visual proof of nature's extreme events, people have become willing to believe that volcanoes are chucking up more land and getting more active. Some scientists have suggested something of the kind. The theory is that the retreat of ice cover depressurizes and thus expands the Earth's magma belt, allowing more molten rock to be created and more eruptions to occur. Writing in *Scientific American*, one of the progenitors of this theory, Dr Graeme Swindles (a physical geographer at the University of Leeds) said, 'I think we can predict we're probably going to see a lot more volcanic activity in areas of the world where glaciers and volcanoes interact.'

It's not a prospect that would impact anywhere near Tonga and, talking to another eminent volcanologist, Dr Nick Cutler whose office is a convenient ten steps from my own at Newcastle University, I hear that the 'increase in volcanism' theory remains controversial. Nick's got some fascinating insights into the global significance of volcanic eruptions. The one that knocked me sideways is that 'the biggest eruption of the twentieth century was Mount Pinatubo in 1991, and the cooling effect of that, over a couple of years, was almost the same as the anthropogenic warming effect over the whole twentieth century.' The 15 million tonnes of sulphur dioxide that this one Philippine volcano released reacted with water in the stratosphere to form particles that scattered and absorbed incoming sunlight, thus cooling the Earth. Drawing on recent ice-core research from Greenland, Nick went on to explain that what he calls the 'volcanic forcing' of global climate change happens far more often than scientists once thought: climate change is 'forced' by 'fairly run-of-the-mill, two or three times a century type eruptions'.

It's a complicating factor: anthropogenic warming of the planet is interacting with a range of currently unpredictable natural processes and events. Nick is a wry observer of the human need to find patterns and predictability in nature. The mismatch between our mayfly-like life spans and geological time makes it hard to grasp that events like new volcanic islands, which seem extraordinary to us, start to look continuous when viewed in geological time. 'In terms of a human lifetime, you may not see magma emitted to make islands,' Nick tells me, 'but across millions of years you'd see them popping up all the time.' We know roughly where new islands will appear on the boundaries between plates (subduction zones) and in places where magma is burning through the crust (hot spots): 'Places like Iceland and particularly Hawaii and other Pacific island chains are going to get continual creation of islands.' But that's as far as he will go: 'When it will happen and how quickly and precisely where – that's much harder to predict.'

Could such island creations have planet-changing consequences? Nick explains that generally, while volcanic activity certainly could and does, island-creation events tend to be less destructive. He has to go back two million years or so to find a counter-example, namely the supervolcano of Toba in Indonesia: 'a sizeable island which then more or less totally blew to pieces' (and today is the site of Lake Toba). This event caused widespread cooling in the atmosphere, and it has been speculated that it nearly extinguished human life on Earth.

For scientists the importance of new volcanic islands is not really to do with their planet-changing potential. It's how they allow us to study, from a blank slate, the formation of soil, of plants, of life. In truth, that is what I am most looking forward to seeing when I finally get to swim ashore and step foot on Hunga Tonga.

Saturday: the day of my journey. Another hot and humid morning. A battered pick-up grinds into view and soon I'm bouncing along Nuku'alofa's seafront with one of Branko's sons, a large, tattooed and quiet young man who has also brought a stack of 'meat sandwiches'. On the dockside Branko, in his trademark and faded 'Drink Beer' t-shirt, is busy and ready. The launch, which is about 12 metres long, has two big outboards; amid their overwhelming din, we are soon thumping straight out, past sand-bars and islets, towards darker water. We're in good spirits and I catch stories of other Tongan islands, such as the mysterious Falcon Island, an ephemeral and actively volcanic island way out to sea that last emerged in 1987. Today it is underwater but it is expected to return. Like the weather, it comes and it goes.

I wedge myself down and cling on. After half an hour I chew down a meat sandwich and risk a peek ahead. I can tell Branko and his son are getting worried – not by the water under us but by what they see ahead. Branko turns and shouts in my ear: 'It's big sea. Doesn't look good.' He points my gaze towards a forward zone in the water, beyond which white caps are lined up in military ranks. I want to keep going: 'Let's just see,' I keep saying. We lunge onwards. At some point, our relationship to the boat changes. Yawning valleys open out between sickeningly high water and the boat rushes down into them. Each time Branko and his son wrestle us out, powering the engines along and up the gentlest slope they can find. But it's getting hard: the outboards shriek indignantly; the boat is beginning to roll and turn on itself. It's time to go.

Back in the harbour Branko is apologetic, his son even more so. They refuse my offer of money for the spent diesel or for the sandwiches. Another time, perhaps. But I guess they know that, for me, this was a once-in-a-lifetime opportunity. I ask Branko to pose for a final photo. He is as accommodating as ever, someone I got to know and like, but his usual jauntiness has collapsed.

At least I have his stories. His adventures since arriving in Tonga at the age of twenty-five could fill any seafaring novel. One I remember concerns the time he sailed to a volcanic island with what he calls 'the Google man – founder and owner of Googles'. With an appalled grin he tells me, 'He came in a private jet; his secretary called me up: "Oh, we want to go the volcano."' A cyclone hampered that trip too. Once they had reached the uninhabited island, the wind picked up and Branko explained they couldn't get back until tomorrow. In response his customer 'picked up a phone and he called Houston. I just looked at him, thinking "who is this guy?"' Branko imitates the commanding boom of 'the Google man': '"Hello, Charlie. Can you give me weather prediction for ..." – and he look at me – "What's our position?"' Houston's advice was to 'get the fuck out of there, there's a cyclone coming'. They hunkered down and made it back the next morning: '8-metre waves, six hours home'.

Whatever the height of the waves or the time taken, islands have a magnetic pull. When they rise up out of the sea, it's an act of creation – at least for land-bound creatures like us. No wonder so many creation myths start with island-building. That's certainly true in Tonga. It's a speckle of low, small islands that are next door to nothing. It's 2000 kilometres to the nearest large land-mass, New Zealand. Stories of island creation are still handed down through the generations. Tongan legend has it that, in the beginning, there was nothing but sea. The ruler of the sky and god of carpenters, Old Tangaloa, became tired of the emptiness beneath him so sent down one of his offspring, in the form of a plover, to see if he could find land. When this mission failed, Old Tangaloa demanded that his son take shavings from the wood carving he was working on and pile them up in the sea. And so the first islands of Tonga were stacked up, the gift of the gods.

The Tongans have also learned that islands come and go. Hunga Tonga burst into the world, and it grew, but one day it will disappear. The sand-bar protecting its crater lake has now returned but it is again being whittled down; once it goes, the ocean will pour in, eating away at the island and hollowing it out, returning Hunga Tonga and Hunga Ha'apai to their former separate selves. But then – it could be tomorrow or in a hundred years – another new volcanic island will just as suddenly cleave the seas. As yet these dramatic births are well beyond our powers of prediction. Volcanic eruptions are hard to call; they may be minor events or be planet-changing – remote and beautiful spectacles or the cause of the end of all life. Our human mastery of the planet, even our much-heralded Anthropocene, is only skin deep.

It's time for my early flight out of Tonga. The black dawn begins to lift, and through the plane window I scan the sea, hoping to catch a glimpse of the island I came so far to see. Maybe I did. There are a few lumps down there, just shavings in the sea. I look again and they have vanished.

THE ACCIDENTAL ISLANDS OF
PEBBLE LAKE, HUNGARY

Why am I hunting for islands in landlocked Hungary? The nation's capital, Budapest, is 542 kilometres from the nearest salt water but I'm intrigued by pictures of a flooded gravel quarry in a southern suburb. It is dotted with islands and has been given the name Pebble Lake. Aerial photographs show the islands have a central green space and a circumference clotted with tiny self-built houses; from the air they look like eccentrically arranged but complete sets of teeth gnashing at the water.

Pebble Lake is an example of a distinct category of artificial island: the accidental island – a side effect of human activity. Like many quarries, this one was only partially dug out. Tall pillars of rock were left standing. It was then abandoned and, because it's deeper than the local water table, it filled with water. The result was a scatter of islands encircled by deep cold water.

The planet's biggest example of an accidental island is René-Levasseur Island. If you type that name into Google Earth you may be in for a surprise. At 72 kilometres wide, it is a very large – and, from high up, very circular – island that looks like a big button dropped on Quebec's northern expanses. One of the biggest objects ever created by *Homo sapiens*, René-Levasseur was an unintended consequence of the flooding of its surrounding land in order to make a reservoir. It is on the opposite end of the scale to Pebble Lake but they are distant cousins and I'm hoping my visit to Kavicsos-tó (the Hungarian for 'Pebble Lake') could be a gateway into this intriguing subspecies of island.

Pebble Lake is a thirty-minute drive from the centre of Budapest. I put the co-ordinates into the satnav and it doesn't take long for the affluent city of palaces and pavement jazz musicians to slide away into a dusty edge-land of grey blocks and car fumes. I steer across an arm of the divided Danube onto Csepel Island. Somewhere at its northern end is Pebble Lake. At 48 kilometres long, Csepel is one of the Danube's largest islands (the biggest is in Slovakia and stretches 84 kilometres) and was once the heartland of Hungarian heavy industry. From what I can see through my increasingly dirty windshield, Csepel is now a post-industrial dumping ground, a sprawling mix of anonymous housing blocks and semi-derelict areas. Perhaps I am prejudiced by the grisly news item I clicked on a few nights before, which named Csepel as the place mobsters and paramilitaries bury their victims alive. In 2011 four bodies were unearthed, all of whom had met this end. It is suspected that the island is the lonely resting place of others. It feels a long way from Budapest's most famous island, Margaret Island, a beautiful park that lies in an affluent reach of the river. I was walking there only yesterday: happy in the hot July sunshine, enjoying the good-natured crowds, the carefully clipped flowerbeds and a delicate cone of lemon sorbet.

Pylons and power lines grid the blue sky as my satnav signals a curt right. It takes me along a straight, potholed dirt road and, with each passing vehicle, clouds of yellow dust choke the breezeless summer air. The radio station is playing 'Hotel California'. I won't be able to shake that tune out of my ears. I'm also remembering from my late-night research that the Pebble Lake community is disputatious and values its privacy. The 166-hectare site (106 hectares of which is water) is privately owned and there is no public right of access. After the quarry was abandoned it started to fill with groundwater and its shores were quickly colonized by weeds, trees and bird life. Within a few years it was a verdant green oasis. Only the fish, of which the lake is famously full, are artificial introductions.

A few turns later and I'm thumping slowly along a twisting peninsula that bends into the heart of the lake. Weekend shacks are irregularly spaced anywhere and everywhere and it feels like I've sneaked into an idiosyncratic and private club. I'm not betting on a friendly invite over to one of the lake's five settled islands. So far, I've seen no one. Seventy families are said to live here year round but, if so, they keep a low profile. The self-built homes are modest but well-cared for in a pleasantly ramshackle way but they all have high fences. It's time to park the car and start walking.

It's mercilessly hot and I'm soon distracted by fat coal-black bees that bounce drunkenly between trumpet-shaped flowers and by a chalk-white moth that makes slow, unsteady steps along my forearm. Near the water's edge I take photographs of the little houses. Some are brightly painted and have manicured gardens and shady porches where rocking chairs wait out the day. Nearly all of them have a small floating pontoon, designed for idling and nursing a cold beer.

My first sight of human life is a tubby swimmer in tight Speedos. Somewhere in his retirement years, he plods heavily down a private jetty 10 metres in front of me and slaps into the water. Suddenly conscious of my unseemly lurking, I make a hasty retreat and scrabble up a steep ridge, past piles of dumped building material, to gain a view over the whole lake. Despite the prosaic names given to them by their residents, such as Bare Island and Small Bald Island, the islands are effortlessly cute. They have no shoreline. There is a vertiginous drop into the depthless waters. However, there are signs of shallow sunken structures that project just under the water – ghostly forms indicating that the waterline has risen since settlement began, covering terraces and pathways. On this high ridge, the grey hinterland of Pebble Lake is visible from every angle. You can hear it too: the grinding of the big city punctuated by periodic shrieks of aeroplanes or train brakes. Amid all this, Pebble Lake looks unreal, as if it has been dropped from another, better planet – a beautiful alien abandoned on a hostile world.

My reverie is terminated by another middle-aged man in Speedos. A more gamey and athletic type, he rushes out from what I took to be an empty house to repeatedly bellow the word 'Finish!' Throughout the years I've spent researching islands, I've been told to 'go away' many times and in many different languages. Waterfront exclusivity is eagerly guarded; it has a hair-trigger intolerance of outsiders. My feelings of admiration flip into resentment at being shut out and shouted at. I'm reminded that Pebble Lake regards itself as a private place, acquired by 664 anglers who joined together to pay the purchase price. Disputes over ownership and access continue to rumble on, with residents siding with one faction or another. On a Budapest news site journalist Janice Kata reports that 'everyone is suspicious of the other'. Talking to 'the elder of the lake system', an octogenarian called József Antal who spends much of the year here, she discovers that thieves have twice ransacked his home; they 'took the gas bottle, even his pile of aluminium cans'. The recent arrival of electricity is heralded by the Pebble Lakers because it has finally allowed them to fulfil a long-held dream: to fit burglar alarms.

Despite its jealously guarded desirability, Pebble Lake was the product of accident not design. Many of the world's 'side-effect islands' share this paradox: unplanned by-products that have come to be seen as desirable addresses. Sometimes the meaning of this value is a source of conflict.

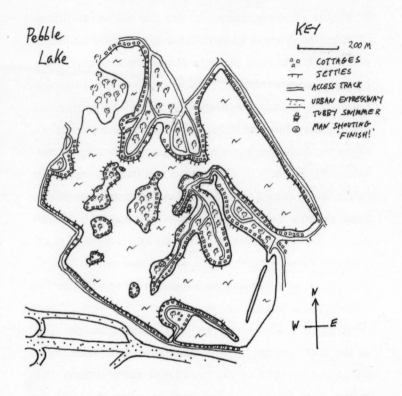

Pebble Lake

KEY
200 M
COTTAGES
JETTIES
ACCESS TRACK
URBAN EXPRESSWAY
TUBBY SWIMMER
MAN SHOUTING 'FINISH!'

N
W — E

René-Levasseur Island is, at one and the same time, a logging resource, a sanctuary for undisturbed nature and a protected slice of native land. It was formed in 1970 when two existing lakes were joined, flooding a continuous loop of land to create a reservoir that is used to power hydroelectric generating stations. The circular shape of both the reservoir and its island is a telltale sign that this landscape was created by a 'falling star'. It was here, 214 million years ago, that a meteorite 5 kilometres long struck the planet. It was the fourth-biggest impact Earth has experienced, forming a crater 100 kilometres wide. What geologists describe as 'post-impact rebound' appears to have been responsible for the centre of the crater rising up to form Mount Babel, the island's 1000-metre mountain.

Although a by-product of Quebec's pursuit of hydroelectricity, the island was soon being thought of as a significant and important place. A portion of the island was set aside for the Louis-Babel Ecological Reserve, and activists began campaigning for the entire island to be protected. In 2003 a coalition called 'Sauvons l'île René-Levasseur' (since renamed 'SOS Levasseur') began lobbying to save what they call 'the integral ecosystem of René-Levasseur Island'. The logging roads that criss-cross the island demonstrate the threat posed and SOS Levasseur's website declares that the island is a threatened natural paradise, home to caribou and golden eagles. Their heartfelt *cri de coeur* is addressed to every 'forest lover and animal protector'. René-Levasseur has no permanent residents, just fifty or so cabins occupied by seasonal hunters. Another 'stakeholder', apart from the logging and mineral rights companies, is the local native population, the Innu. They too have been campaigning to stop all logging activities on the island, which is part of their ancestral land. The campaigners make similar claims for the great forests that extend in all directions around René-Levasseur but the island has come to provide a focus for these concerns *because* it is an island and, hence, special.

A similar story can be told of the islands formed at the start of the last century, thousands of kilometres to the south of René-Levasseur, when Gatun Lake was created in Panama. A network of flooded valleys, the lake is a key component of the Panama Canal, providing headwater to fill the locks. Every ship that transits the canal uses 202,000 cubic metres of water from the lake – water that then flows out into the Atlantic or Pacific. Over the years, the verdant, tropical islands created as a by-product of the creation of Gatun Lake have come to be valued as wildlife sanctuaries and eco-tourist attractions. The world's most famous and oldest artificial 'eco-island' is here: Barro Colorado Island, which was established as a nature reserve in 1923. The island is home to the Smithsonian Tropical Research Institute and is considered one of the few places on the planet where an untouched tropical ecosystem can be studied.

Modern industry requires continuous gouging of the Earth's surface: mining, drilling, building, dredging. A lot of debris is created and it has to go somewhere. Often it gets washed downstream. In some places the accidental islands that result are places of reprieve, the silver lining in an age of ugliness. This redemptive quality is certainly to the fore in Florida's Spoil Island Project. If you look closely at satellite images of the east coast of Florida you'll see them dotting coastal lagoons. There are 137 spoil islands in the Spoil Island Project: industrial by-products that over the past few decades have been reclaimed as environmental reserves and eco-tourist destinations. Colonized by sea grasses and mangroves, the islands are managed by Florida's Department of Environmental Protection. They designate some as 'conservation islands' and others as 'recreation islands', with the latter offering very basic camping sites with picnic tables and fire rings. The Florida Spoil Islands are an encouraging example of how by-products of human industry do not always have to be places of abjection. For some, even the words 'spoil island' may send a shudder down the spine. They could have been relabelled and a lie told – branded as 'paradise beaches' – but I think the Florida project has got it absolutely right: they call them what they are.

Despite having been told to 'Finish!', I'm still mooching about the byways of this densely settled former quarry, enjoying some of the quirky 'art' its residents have assembled from advertising hoardings and assorted junk. It's a hive of private individuality. It has certainly satisfied my feeling that the lure of islands is not confined to countries with shorelines. 'Landlocked' is an unfortunate, graceless kind of label. There is no more reason to cast 'landlocked' Hungary into a dungeon of topographic isolation than salty, seafaring, 'sea-locked' Britain. In Hungary too, in a low-key kind of way, it is the age of islands.

I'm back in my mud-splattered hire car; it's time to leave this green refuge. On the interminable dirt road out of Pebble Lake I edge past yet another middle-aged man wearing only a small pair of Speedos and flip-flops. It must be a Hungarian thing. He has a long walk in front of him; perhaps I should give him a lift. I slow down but he curtly waves me on; like other Pebble Lakers, he is gruffly self-sufficient. As with other private islands – accidental or not, owned by millionaires or, like these, by ordinary people – the default gesture is a wave of the hand, not as a signal of welcome but to tell the outside world to go away.

TRASH ISLANDS

Deep into the future, a geologist traces a finger along a thin, darkly resinous deposit in the strata and, before moving on to more substantial layers, mumbles something about 'the plastic age'. It's not an implausible vision. A defining feature of the modern world is its waste: rubbish is being produced in such abundance and from materials of such durability that it is forcing us to rethink geology. I want to take another step in this conceptual revolution and make the case that it is also changing the way we think about islands. Trash islands form a global archipelago that reaches from the smallest floating clot far upstream to the unimaginable expanses of the Pacific Trash Vortex.

One hot morning a few years ago I took a long and bumpy taxi ride to visit Cairo's 'Garbage City'. The Coptic Christians who call it home make a meagre living by recycling as much of Cairo's rubbish as they can, which is pulled in by donkey carts and little vans and sorted in noisy household workshops into different types of metals, plastics and paper. Six hours later I was downtown: invited by a wealthy Cairo resident, a friend of a friend, to a swanky bar high above the Nile. As we walked on the rooftop terrace, admiring the glitzy panorama as ice cubes jingled in our G&Ts, I tried to make a joke about being a million miles from 'Garbage City'. My new pal grinned sardonically and pointed down to the river. In small boats would-be fishermen were pushing their way through dense mats of floating trash. Elegant white egrets picked their way across the ugly surface as the plastic waste sealed up each boat's wake.

Plastic takes between 500 and 1000 years to degrade and we make hundreds of millions of tonnes of it every year. Every decade production of plastic more than doubles. Little of it is being recycled; most is burned, buried or just tipped into rivers, eventually washing far out to sea. The thickest and most common trash islands can be seen along coasts and in rivers in Africa, Asia and much of the Americas. Recently news websites like Newsflare, which allow anyone with a smartphone to upload video of local stories, have provided the most urgent and grim testimony of the ubiquity of trash islands. Filmed in October 2107: 'Shocking scenes of vast quantities of plastic waste floating along a river in Mexico's southern Chiapas state'. Filmed in September 2018: 'Rivers of plastic flowing in the [Spanish] province of Almería'.

Filmed in January 2018 in Bukittinggi, Indonesia: 'Water for human living and irrigation has been polluted by plastics and rubbish'. In some of these rivers, fishing has become impossible and what fishermen are left have swapped profession, turning their boats into trash-breakers and thrusting through the flotsam to look for saleable waste.

Once offshore, trash islands usually break up, which makes clearing up their toxic cargo all the more difficult. It is only when they stick together – because of some whim or habit of the current – that people take much notice. The 'garbage island', about a kilometre long and weighing about 100 tonnes, that formed in the Gulf of Thailand in 2017 was soon spotted and branded as an unsightly disgrace by the local media. Swift action followed: speedboats with fishing nets were sent out to scoop it up. Unfortunately, this kind of prompt response is the exception rather than the rule. Even when it does happen, much of the debris will already have been missed. Plastics that are dispersed, broken up and sunk are out of sight and out of mind. Many of the commonest plastics, such as Polyethylene terephthalate (PET, which is used to make drinks bottles) and polyester, are relatively heavy and quickly sink.

Another thick plastic agglomeration that occurred in 2017, this time off the coast of Honduras, shows how dealing with the problem, even when it is within easy reach, is rarely straightforward. The mayor of Omoa, the nearest town, complained that the clean-up effort was beyond his resources: 'On Friday, we filled twenty dump trucks of thirteen cubic meters each, and it made almost no difference.' The Hondurans claim, not unreasonably, that the waste was created upstream. They blame the Guatemalans. The local tourism chief drove the point home by taking journalists on a tour to show off the Guatemalan labels on the plastic bottles scooped from the sea.

Responsibility for the pollution of water can be impossible to pin down, with different regional and national authorities all pointing at each other and arguing that the real source of the problem is further upriver. The biggest rivers are the biggest carriers of waste and they often traverse a number of countries. It is now thought that much of the plastic waste found in the world's oceans comes from just ten rivers, eight of which are in Asia. One of them is the Mekong, which travels through China, Myanmar, Thailand, Laos and Cambodia before disgorging its load in Vietnam – at which point blame has become almost as thoroughly dispersed as the plastic itself.

Some of the plastic that feeds the garbage patches forming in the middle of the large-scale circulation patterns, or gyres, of our oceans has been thrown off ships but most has arrived from beaches or rivers. Footballs, kayaks and Lego blocks have all been spotted, along with the usual mass of plastic bottles and fishing net. Most of the plastic has broken up into fragments. The oceans' garbage patches do not exist as a single entity but a soup or galaxy of rubbish, most of which has sunk to the ocean floor or floats just below the top and sometimes gloops together on the surface. Oceanographer Curtis Ebbesmeyer has argued that these patches 'move around like a big animal without a leash', and every so often they find a shore and cough up plastic all over the beach. Ebbesmeyer puts it in suitably grotesque language: 'The garbage patch barfs, and you get a beach covered with this confetti of plastic.'

All oceans have circulating currents and, since rubbish is being picked up by such currents around the world, so trash vortices are forming in every ocean. In fact, the Pacific has two: an Eastern and Western Patch. Intimations of a North Atlantic Garbage Patch came in 1972 when oceanographers discovered plastic pieces in the Sargasso Sea. The North Atlantic's debris zone moves seasonally, drifting 1600 kilometres north and south each year. Photos of the North Atlantic Patch, taken by the research vessel *Sea Dragon*,

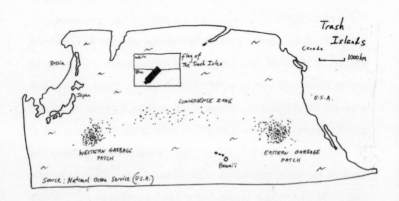

which since 2010 has been studying trash vortices all over the globe, show a mass of floating debris on rough sea.

Estimates of the size of the oceans' various trash vortices vary hugely. The Pacific Trash Vortex, which is the one that gets all the attention, has been measured as 670,000 square kilometres but also as 1,500,000 square kilometres. However large it is, it is clearly highly contaminated. It has been estimated that at its centre every square kilometre contains 480,000 pieces of plastic. Animals are ingesting this fragmenting debris and so taking in toxic pollutants. In one 2018 study, marine scientists from the National University of Ireland found plastic in 73 per cent of the deep-sea fish collected from the Atlantic. It's not just marine life that is impacted by the dispersal of plastic; a review conducted by the World Health Organization in 2018 found microplastics in 90 per cent of bottled water.

The Pacific Vortex was discovered by yachtsman Charles Moore in 1997 on his way back to Los Angeles from Hawaii. He decided to take his yacht into a part of the ocean usually avoided by sailors because of its slow currents and lack of wind. To his astonishment, he found himself sailing into a sea of gunk: 'Every time I came on deck, there was trash floating by.' Moore, who has since devoted himself to cleaning the oceans, says of the Pacific Garbage Patch that it is 'just absolutely gross – a truly disgusting plastic cesspool' and that it 'has to be burned into the consciousness of humanity that the ocean is now a plastic wasteland'.

More recently, some scientists have been using the term 'plastisphere' to think about the new kind of ecosystem that all this debris is creating. Plastic is quickly colonized by a diverse array of microbial life. One of the marine biologists who came up with the term, Linda Amaral-Zettler of the Royal Netherlands Institute for Sea Research, points out that the plastisphere 'is really quite a little zoo'. Larger organisms are also making use of marine plastic, such as marine worms who researchers at the Korea Institute of Ocean Science and Technology have discovered are eating plastic and excreting microplastics.

More recently, scientists have been taking a sceptical tone about 'trash islands'. In a piece published on the science blog io9 titled 'Lies You've Been Told About the Pacific Garbage Patch', tech journalist Annalee Newitz lays into the 'myth' that 'There is a giant island of solid garbage floating in the Pacific.' Another counter-blast came in 'The Dirt on Ocean Garbage Patches', published in *Science*, where Jocelyn Kaiser tells us that 'Their biological impact is uncertain and their makeup, misunderstood.' This myth-busting shtick makes for arresting headlines but rather misses the point. Of course the Pacific Garbage Patch is not a solid island where you can build a house or park your car. No serious report on it has ever suggested anything of the kind. The Pacific Garbage Patch is one end of a spectrum, along which a diverse flotilla of trash islands can be arranged – from the soup-like to the immobile and metres thick. Each is an evolving, changing form; even the most stuck and solid-seeming river trash eventually gets washed downstream, while the liquid ocean 'vortices' will get thicker and more lumpy as more ingredients are poured into them.

Much of the impetus for thinking about the Pacific Garbage Patch as a new island is rooted in environmentalists trying to make people care. They are asking us to shift our ideas. It's a geographical reboot, a necessary shake-up. The plan to ask the UN to accept the Trash Isles as a new country was dreamed up by two environmentalists and advertising creatives, Michael Hughes and Dalatando Almeida. Talking to an advertising trade magazine, *Creative Review*, they explain: 'We wanted to come up with a way to ensure world leaders can't ignore it anymore, a way to stick it under their noses, literally.' They were struck by the fact that the Pacific Garbage Patch covers a country-sized area, and so 'With no one paying attention to this catastrophe' they submitted a Declaration of Independence to the United Nations.

If we become a country and a member of the UN, we are protected by the UN's Environmental Charters, which state ... 'All members shall co-operate in a spirit of global partnership to conserve, protect and restore the health and integrity of the earth's ecosystem'. Which in a nutshell means that by becoming a country, other countries are obliged to clean us up.

The Trash Isles campaign has been slick but playful, with the would-be country's flag, passports, official stamps and a currency (twenty-, fifty- and hundred-Debris notes) all splashed with artful portraits of both the rubbish and its victims. On the twenty-Debris note there is a turtle, its waist girdled by plastic. The Trash Isles established a monarchy and appointed Dame Judi Dench as its queen. At the last count 132,000 people have asked to be citizens, the first being former US vice president Al Gore. A spokesman for the UN's Secretary General declared that the Trash Isles campaign was 'creative and innovative. But the chances of it being accepted are fairly nil.'

The Trash Isles bid to the UN was prefigured a few years earlier by the Italian artist Maria Cristina Finucci, whose Federal State of Garbage Patch was declared to be a nation on 11 April 2013 at UNESCO Headquarters in Paris. This rival state's territorial claim encompasses five garbage patches as a 'federal state with a "population" of 36,939 tons of garbage' and a total area of 15,915,933 square kilometres.

These two nation-making projects have overlapping territorial claims but appear to be ignorant of each other's existence. Our polluted seas are provoking a growing community of artist-activists, with much of the resultant work playing with images of islands or going one step further and actually building islands out of plastic. The éminence grise of such enterprises is the British-born eco-artist Richart Sowa, who has been building and living on tiny islands made of plastic bottles off the Mexican coast since 1997. The latest version, Joysxee island, is 25 metres wide and 30 metres long, floating on approximately 150,000 plastic bottles. On his website Sowa says: 'I AM LIVING on a Franchiseable Prototype FLOATING ECO/ISLAND which SOLVES the PROBLEMS of,,,, INCREASING TRASH, by preserving it in net bags under used shipping palettes to provide a floating form for humans, animals, marine life and a garden to flourish on.'

In 2018 Rotterdam-based Recycled Island Foundation launched their floating Recycled Park. Plastic waste recovered from the busy harbour has been shaped into a series of hexagonal platforms that are used as refuges for plants and small animals. It's a modest endeavour but points the way. After all, just clearing up the plastic is only part of the problem; we also need to figure out what to do with it. Another Dutch initiative is scaling up Rotterdam's idea. Dutch company WHIM Architecture has outlined a plan for a floating city of half a million people living on an island made up of waste recycled from the Pacific Garbage Patch. 'The proposal has three main aims,' they declare, 'cleaning our oceans from a gigantic amount of plastic waste, creating new land and constructing a sustainable habitat.'

In 2018 yet another Dutch scheme, Ocean Cleanup, prototyped a marine 'sweep', a long boom that captures refuse which is then picked by boat. Ocean Cleanup claim that 'A full-scale system roll-out could clean up 50% of the Great

Pacific Garbage Patch in just five years.' Others have been less impressed, warning that the system will catch only the most visible fraction of plastic pollution while killing creatures such as turtles and floating plankton. Another scheme that is attracting attention is the deployment of plastic-eating bacteria. A Japanese research team sifted through hundreds of samples of discarded PET plastic before finding a colony of organisms using the plastic as a food source. But the researchers are cautious: Professor Kenji Miyamoto from Keio University noted that there are 'many issues' still unresolved and that 'it takes a long time'. Since this bacteria only works on PET plastics, which are already 100 per cent recyclable, it is unlikely to be the breakthrough that is desperately hoped for.

It's a modern conceit to imagine that – having filled our rivers, oceans and indeed our lives with plastic junk – some clever scientist will invent a gizmo and clear it all away. It allows us to sidestep the fact that the scale and spread of plastic entering our seas and rivers – and hence the scale and spread of trash islands – is increasing at an alarming pace. It's this that must be tackled.

Deep into the future, the geologist has made an exciting find. All that remains from the plastic age is now reduced to microparticles, a permanent geological feature since it cannot be broken down by micro-organisms. But every blue moon you can strike lucky. And here, unbelievably, revealed from deep history, is a fragile bluish fragment of plastic fabric hanging delicately from a crumbly black layer. What a find! A museum piece, of course – a rare relic from a period in which (so the current theory goes) *Homo sapiens* both drank and swam in solutions of toxic plastic because they thought it would cure their ailments and stave off death. That's the current theory; in truth, no one is at all sure what they were thinking.

Early growth and nesting birds on Hunga Tonga-Hunga Ha'api (*Photograph by Branko Sugar*).

Branko back in
port after our
unsuccessful trip.

Pebble lake, Budapest. Small floating pontoon, designed for idling and nursing a cold beer.

One of the islands of Pebble Lake.

San Blas Islands. A typical small inhabited island.

A typical San
Blas shoreline.

Tupsuit Dummat. *Ulu* canoe and artificial island just offshore.

Toilets poking out from Tupsuit Dummat.

Bernado building up his artificial island, Tupsuit Dummat.

Fafa, Tonga. Its southern shore is an obstacle course of palms and wooden posts felled by the surging waters.

Tongatapu, Tonga. Boulder sea defences on the north coast.

Storm damaged house, Tongatapu, Tonga.

Rising seas in
Hugh Town
harbour, St Mary's.

Iron Age village of Halangy, St Mary's.

Aphra on top of a coastal burial cairn, St Mary's.

Another, even smaller, unnamed island in Loch Awe.

PART TWO

DISAPPEARING

Disappearing Islands

DISAPPEARING ISLANDS ARE the star exhibit of sea-level rise. This is perverse. Most of the world's coastlines are at risk from sea-level rise. By far the greatest impact will be on low-lying coastal plains crowded with cities, where hundreds of millions of people now live. Along the eastern seaboard of the USA, or the densely settled coasts of South East Asia, surging waters will create more islands than they destroy, turning landscapes of valleys and hills into archipelagos.

Yet the image of the inundated island remains compelling. And urgent: they are disappearing right now, especially low-lying islands in the tropics. What is happening to them is an early warning to the rest of the planet. There are less rational reasons why vanishing islands touch us. To see a whole place – often an ancient one where people have lived for hundreds or even thousands of years – erased from the horizon is painful in a way that, for the moment, forecasts of the flooding and subsequent 'break-up' of continental seaboards are not.

On the next part of my island adventures I will travel to islands threatened by sea-level rise. The three island groups I will be exploring could scarcely be less similar. They are certainly far apart: one lot are in the Caribbean (the San Blas Islands of Guna Yala), another in the North Atlantic (the Isles of Scilly), and the third are in the South Pacific (Tongatapu and Fafa).

Like a number of other low-lying tropical islands, the San Blas, which belong to Panama, are facing imminent evacuation. That is not true of Fafa and Tongatapu, which are in Tonga. Not yet, although many problems are being thrown at this redoubtable kingdom: it is facing environmental, social and economic crises that interlink and feed on each other. England's Isles of Scilly are, by contrast, affluent and appear very content. But the sea is a great leveller and the sea is rising here too. The Isles of Scilly – the most westerly point of England – intrigue me in other ways. For you can see on many of them clear evidence that sea level has been rising here for a very long time.

These three island groups don't exhaust the ways that islands vanish. Islands subside and erode, and volcanic and tectonic forces can destroy islands just as easily as they create them. There also is a darker story to tell about the death of islands. Intense forms of human exploitation, such as mining or nuclear testing, have left the world with a sad profusion of destroyed islands.

Rising seas

Small flat islands, especially those in warmer latitudes, are very vulnerable to sea-level rise. It's odd, then, that building small flat islands in warmer latitudes is such big business. One day the dots will join. Without the installation of pumping equipment and stout walls, many of our 'age of islands' newest creations will not have a long life.

News stories about disappearing islands can be alarmingly short-sighted. Sometimes tagged as 'breath-taking places to visit before they disappear', they often leave the impression that sea-level rise is a problem confined to remote, unpronounceable, Pacific atolls. Hand-wringing about 'the plight of the Kiribatians' translates a global, universal crisis into something comfortingly far away, affecting a reassuringly tiny group of people.

The Pacific *has* seen more than its fair share of losses, however. Since 2007 the islands of Laiap, Nahtik, Ros, Kepidau en Pehleng and Nahlapenlohd, which were all once part of the Federated States of Micronesia, have gone. Five of the Solomon Islands have also been lost. Unlike most island disappearances, these losses have been widely reported, often accompanied by a link to other stories on places that will go the same way. The Maldives, Palau, Fiji, Tuvalu, Seychelles, Kiribati, the Cook Islands and French Polynesia all regularly appear on lists of vanishing countries.

It is revealing that the journalist I cribbed that roll-call from finishes her article with an incredulous footnote: 'Even the United States is affected by rising sea levels.' The idea that sea-level rise impacts *us* has not quite sunk in. She cites the best-known American example: the islands of Chesapeake Bay, an estuary that separates Maryland and Virginia. But let's not get too worried. President Trump called the mayor of one of the Chesapeake islands, Tangier, telling him he shouldn't worry about sea-level rise – it is, Trump says, fake news. Those looking for a more sober witness can find it in William Cronin's island biography, *The Disappearing Islands of the Chesapeake*.

Globally, sea and air temperatures are higher today than at any time since records began. However, it is not getting hotter at the same rate everywhere and nor are sea levels rising uniformly. The rate of rise is determined by a complex mixture of sea currents, the gravitational pull of the polar ice sheets (the thinner they are, the less they pull water away from lower latitudes), the fact that warm water expands, post-glacial 'bounce-back', and local factors such as land subsidence. Even within the same country there is huge variation. In the Philippines the sea is rising 14 millimetres each year in Manila, the country's capital, but go a few hundred kilometres to the south, on the island of Cebu, and the rise is much less: 0.9 millimetres per year.

The range of predicted rise, averaged across the world, is also considerable, from just 26 centimetres to nearly 3 metres by the end of the century. The estimates vary but all these predictions point to an accelerating trend: increases that are now 'locked in' will continue to grow. Moreover, these predictions do not take threshold breaks into consideration; tipping points when what we have seen so far (slow-building acceleration) will shift into another gear. One of these thresholds is the point at which the polar regions get too warm to sustain their ice sheets. The ice sheets that cover Greenland and the Antarctic hold enough water to raise sea levels by about 65 metres, which is the height of the Sydney Opera House. The important point is that ice-sheet melt has not *yet* made a substantial contribution to sea-level rise. Most of the rise we have experienced has been from the thermal expansion of the sea, with a smaller portion due to melting glaciers. If temperatures continue to rise, the ice sheets *will* melt in a big way and that will change everything. In early 2019 the ominous news came that the Greenland ice sheet is melting four times faster than previously thought. 'This is going to cause additional sea-level rise,' says the report's lead author, Michael Bevis, a Professor of Geodynamics at Ohio State University, adding, 'We are watching the ice sheet hit a tipping point.' Bevis is sombre: 'The only thing we can do is adapt and mitigate further warming – it's too late for there to be no effect.'

We are headed into the unknown but not the unfore-warned. Some, but not all, disappearing islands can be defended. We should not assume that to 'adapt and mitigate' always means having to evacuate or that what we will be witnessing over the coming centuries will be a straightforward universal drowning. Just as islands can be artificially built, so they can be artificially protected – sometimes aided by nature herself. In a recent study of Tuvalu's atolls and reef islands, Professor Paul Kench of the University of Auckland found plenty of evidence of sea-level rise but also of island growth caused by sediment deposited by storms. Overall Tuvalu's total land area actually got bigger between 1971 and 2014 by 2.9 per cent. It's not much of a counter-trend, however, and bulked-out beaches will be no match against remorseless sea-level rise. But Kench is right to counsel against the idea that, at least in the short term, the disappearance of low-lying islands is inevitable and that they should all be abandoned. The crisis is a real one and the long-term outlook is not good, but the complexity of the processes reshaping islands coupled with human ingenuity suggests that mass relocation should not be the first solution we consider.

Eroding and exploding islands

The island of Esanbe Hanakita Kojima was only named in 2014 and the next time people looked for it, off the far north shore of Japan, it was gone. The Japanese government named it, along with 158 other uninhabited islands, in order to shore up territorial claims on its northern seas.

But the stormy wind and ice flows that barrel along this forbidding stretch of coast had other ideas and, after scraping and blowing away the island's surface, they removed it from the map. These are natural processes that, many times every year, remove islands. Esanbe Hanakita Kojima went under in 2018, the same year that East Island in Hawaii was reported missing. At 800 metres long and 120 metres wide, East Island was a wildlife haven for seals, turtles and albatrosses but a hurricane rattled by and it was dashed to bits.

Separating the natural from the unnatural is getting tricky. Since global warming is making storms more frequent and harsher, more islands are probably being washed away. Ice melt in the high north also makes islands more vulnerable. In 2008 the population of the Alaskan island of Kivalina, who will soon need to start new lives on the mainland, filed a lawsuit against ExxonMobil. They were suing for their relocation costs but the case was thrown out of court on the basis that doing something about greenhouse gases was not the responsibility of oil companies.

The interaction of the natural and the unnatural can also be seen in land subsidence. Along many coasts the land is falling at a faster rate than the sea is rising. It's a particular problem in South Asian cities, like Jakarta, Ho Chi Minh City and Bangkok, where huge quantities of water for human and farm use have been pumped out of the ground. The impact of subsidence on islands is especially alarming in deltas because so many people live on them: about

500 million people today. But they are sinking fast. Three-quarters of the Ganges delta – home to 130 million people – faces inundation. China's Yellow River delta is currently dropping so fast that local sea levels are rising by up to 25 centimetres per year. In some places, the Mekong River delta has been found to be falling by 70 centimetres a year and it is projected to carry on falling by 80 to 150 centimetres within the next two decades. These are colossal numbers and, while some of the causes are purely natural, some are not. Groundwater extraction – much of it to feed the aquafarming of fish, shellfish and suchlike – acerbates subsidence. The coast is also threatened by cutting down coastal mangrove forests and upstream river damming. Dams have all but halted the arrival of new sediment to the deltas of the Nile, Indus and Yellow rivers. This means new islands are not formed and old ones are washed away by floods. Although in some places, like Bangladesh, polder-building is helping to create new farming islands, the prospects of hundreds of millions of delta-dwellers are not bright.

Volcanoes and earthquakes can destroy islands as well as create them. Two-thirds of Krakatoa Island was obliterated by the 1883 eruption. The explosion was heard 4800 kilometres away, and killed about 37,000 people, and the amount of ash released caused global temperatures to fall by as much as 1.2 degrees Celsius. The death of islands can be world-changing. But islands have a rhythm; they come and go and come again. The curious story of Graham Island, which lies in a region of underwater volcanoes between Sicily and Tunisia, is a case in point. After a series of earthquakes, it was spotted on 19 July 1831 and described by a British naval officer as 'a small hillock of a dark colour a few feet above the sea'. Within a month the island was 65 metres high and had a circumference of 3.5 kilometres. Like flies to honey, competing territorial claims soon descended on this grey heap. A Union Jack was planted on 2 August 1831 and the island named after Sir James Graham, First Lord of the Admiralty. It was then named Ferdinandea and claimed on behalf of Ferdinand II by the Kingdom of the Two Sicilies, recently formed by the unification of the Kingdom of Sicily and the Kingdom of Naples. Then the French tricolour was raised on the island. Its French discoverer named the island Julia, in honour of the month it first appeared. However, like most new volcanic islands, Julia was made of ash and light rock and quickly began to disappear. By December the island was nothing but a low reef. The Italian volcanologist and Catholic priest Giuseppe Mercalli wryly observed, 'All that remained of Julia Island were the many names imposed upon it by travellers of various nations who had the good fortune to witness the spectacle of its formation and disappearance.' The story is, of course, not over. The island has been active again, and in 2002 Italian divers planted an Italian flag on the top of its underwater summit, getting in their nation's claim in good time before Graham/Ferdinandea/Julia's next appearance.

Islands we have destroyed

Arguably the world's most frightening island is Runit Island. It is a tiny Pacific island inhabited by a monster. A vast concrete dome called 'The Tomb' sits at one end, packed with nuclear waste and occupying about a quarter of the whole island. Many of the Marshall Islands became uninhabitable, including Runit, after the US conducted sixty-seven nuclear tests across the archipelago between 1946 and 1958. The Tomb was supposed to be part of the solution; it was built in 1979 as a repository for 73,000 cubic metres of radioactive waste. Unfortunately, rising sea levels mean that water has leaked into the dome. A 2013 US Department of Energy report on Runit is a litany of alarming descriptions of deformations, cracks and leaks in the concrete shell.

The Marshall Islands is also where we find the most famous casualty of the USA's nuclear testing programme, Bikini Atoll. Some of the devices tested were colossal: one bomb dropped in 1954 had an impact 1100 times larger than the one dropped on Hiroshima. During these years, Runit and Bikini accounted for more than 50 per cent of the world's nuclear fallout.

Islands that have been used as nuclear test sites are, of course, best avoided: the water cannot be drunk, seafood cannot be eaten, and plants cannot be farmed. But small islands have also been the site of biological weapons tests. Gruinard Island off the west coast of Scotland became 'Anthrax island' when, in 1942, Britain started to test anthrax on the island's sheep. Gruinard was decontaminated in 1990. The Soviet Union constructed a biological weapons test site, called Aralsk-7, in 1954 on Vozrozhdeniya Island ('Renaissance Island') and neighbouring Komsomolskiy Island in the Aral Sea. The islands were also used to develop new bioweapons that were reportedly tested on prisoners. Dubbed the 'Island of Death', Aralsk-7 saw a series of accidents and scandals, such as an incident in 1971 during which smallpox was accidentally released, killing at least three people. Abandoned in 1991, the islands continue to store containers full of an unknown variety of biological weapons. The containers are not maintained and it is said they are leaking. Since irrigation schemes have drained the Aral Sea, turning Vozrozhdeniya Island into a peninsula, the worry is that whatever bioweapons were developed there may find their way out to the wider world.

Another form of destruction inflicted on islands is mining. One of the most spectacular examples is Japan's Hashima Island ('Battleship Island'), which in 1887 began to be covered with buildings that eventually grew into high-rises housing thousands of coal miners and their families. By 1974 the undersea coal seams were exhausted and Hashima was abandoned, though it has had an interesting afterlife. Resembling a collapsing industrial castle, Hashima has a dystopian appeal and today attracts filmmakers and tourists.

This unlikely upside of despoliation will not save Nauru, a 21-square-kilometre nation in the South Pacific once covered in valuable phosphate deposits. Over the last hundred years the entire island has been strip-mined – first by European powers and then, since 1968, by its own government. Once one of the world's wealthiest nations, Nauru is now a wasteland where little can grow. Today its few sources of income include selling passports to foreigners and hosting refugees refused by other countries.

These islands have been destroyed by human activity but they are still on the map. Over the past two decades numerous islands were completely eradicated by mining. The global demand for sand, fuelled by Asia's building boom, is the main cause. Journalist Vince Beiser, who has spent years investigating the global sand trade, reports that in Indonesia alone about two dozen small sandy islands have been dug out and are now below sea level.

Islands are under threat from the South Pacific to the North Atlantic. For islanders these are anxious times.

THE SAN BLAS ISLANDS OF
GUNA YALA, PANAMA

Justino chops a leathery hand against one knee: 'In December the water comes here.' He is a Kuna, one of Panama's indigenous communities. Although he's grinning, worry lines bind his forehead as he tells me about sea rise on his home island. It is one of about fifty inhabited by the Kuna in the San Blas archipelago of Guna Yala ('Kuna land'). The whole group consists of 365 islands and stretches for over 200 kilometres on the Caribbean side of Panama.

We're 70 kilometres but a planet away from the high-rises of Panama City. The Kuna are poor and their islands are minuscule – often just a football pitch in size and a few centimetres above sea level. From the shore or sailing past, all you see of many of the islands are some palm trees and a few huts fashioned from palm fronds. The Kuna came to live on these scattered sandy islets about two hundred years ago. The islands were free of the insects and wild animals that plagued – and still plague – the thickly forested mainland. They also offered protection from other hostile tribes. The islands were sanctuaries: with few pests, surrounded by fish, crabs and lobsters but close enough to the mainland to have easy access to farmland, firewood and fresh water.

Justino tells his story in the old but spotless 40-foot yacht on which I have rented a berth and which is captained by his friend, a softly spoken sixty-year-old: a Majorca-born mariner called Toni. This archipelago and this cramped boat are Toni's home and he is keen to point out the islands that have disappeared. 'Over there, that was an island,' or: 'We are passing straight over another island,' or: 'Again, again, all gone.' In every direction Toni jabs his finger at yesterday's islands. What remains are shallows, sometimes marked by a few sticks and a cresting wave.

A battered copy of *The Panama Cruising Guide* by Eric Bauhaus lies below deck. Its detailed maps make it indispensable for the small yachting community who anchor in convivial packs. Toni's copy is so well thumbed that it is now a spineless wad of weathered pages. During my few days in San Blas, and with little else to do once it got dark, I transcribed its maps and local lore. Bauhaus is clear: 'Every time I do a survey after being away for some time, I have to take islands off the maps that are now nothing but shoals.'

Justino gestures to the mainland: a green mass, alive with mosquitoes and crawling with snakes and not a few crocodiles, just a kilometre or so away: 'We have to go there,' he says, running his hands through his hair. 'My head whirls with problems.' He's expecting that, sometime soon, everyone on the seven islands that make up the Robeson group where he now lives will have to relocate. The Robesons lie at the northernmost fringe of San Blas and are so remote that they don't even show up on Google Earth. It seems that different island groups will head off in different directions and see what happens: 'They go there; we go here – all different.' Justino's smile is fading again. It's a chaotic scenario and what the future will bring is uncertain for him and his extended family. But on only one point he is clear: 'We leave, all go.'

There have been fitful plans for an ordered evacuation. I have brought with me a report by an NGO called Displacement Solutions, which runs through the travails of a central government-funded scheme for the relocation of the relatively large and densely settled Kuna island that everyone here calls Carti. The rough track to the San Blas coast takes you past the empty, half-finished and overgrown accommodation blocks that one day, supposedly, are still going to provide homes for the Carti islanders. Government funding was pulled some time ago and the 'relocation village' is disappearing into the jungle. At the moment it looks like the Kuna's 'displacement solutions' will be ad hoc and unassisted.

The San Blas islands continue to be presented in tourist brochures as shimmering flecks of paradise. On the surface, it is easy to see why. They are tropical, palm-fringed and flung across a warm and usually calm sea. The wide availability of cannabis and cocaine (I am told cocaine on the islands is cheaper than Coca-Cola) adds to the intoxicating atmosphere and pulls in Panamanian partiers and backpackers.

A few Kuna actively provide for the tourist drug trade but others scorn it. Separate islands are under the control of elders called Sahilas and it is they who decide what is permissible. The Robesons are one of the more traditional areas and even beer is banned. Justino lives on the most populated of the Islas Robeson, Tupsuit Dummat. It is a hugger-mugger labyrinth of sandy lanes and running, laughing children. It's happy and well cared for. Weaving around the palm-leaf huts, I pass a neatly painted blue school and the dark entrance of a stoutly built 'congress hall', where the Sahilas issue their edicts, traditionally by singing long sacred songs. Over on Carti, which is regularly overrun with visitors and more plugged in to the modern world, modern manners prevail: the locals tend to be brisk, unsmiling and commercially savvy. But on Tupsuit Dummat there is warmth in the smiles and a sense of excitement that visitors have come. Like nearly all Kuna women, those on Tupsuit Dummat are dressed in colourful handmade shawls decorated with 'mola' – complex embroidery designs that are Panama's most famous handicraft – and plenty of leg bangles. Some also have golden nose plugs and henna tattoos on their faces. Offshore, men and women can be seen paddling in the Kuna's elegant dug-out canoes called *ulus*, bringing sugar cane and fruit as well as firewood and water from the mainland.

It's an idyllic scene but there is a sense of uncertainty in every conversation. Life is getting harder. Interviews collected by Greta Rybus, an American photojournalist, on a Kuna island called Coetupo about 160 kilometres south of the Robesons provide a vivid portrait of the many problems people face. With climate change have come higher tides but also warmer waters and this has led to a decline in the fish catch. 'Now the sea can't heal the way it used to,' a

Coetupo elder told Rybus, explaining that, 'It is too hot [so] nowadays there is not much fish. Before, there were a lot of coconuts and bananas. But not now, because of the changes with the sun.' A teacher who lives on Coetupo described how some of the measures the islanders have used to protect their shore have only made matters worse. 'When people realized the sea level was rising,' the teacher told Greta Rybus, 'they started to destroy and use corals to build a kind of wall, which is very damaging.' This local educator went on to identify one of the key problems with ad hoc, unco-ordinated relocation plans: 'About five years ago, we tried to start a project to relocate to the mainland, but there were disputes within the community. The land on the continent is already distributed; it has owners. And these owners don't want to give their land to other people.'

Back on Tupsuit Dummat, I've been invited inside Justino's family hut. Thin sunshine patterns the mud floor with slips of yellow light, pitted and smoothed by last winter's flood. Occupied hammocks swing from every roof beam. At the back is Justino's grandmother, too sick to raise her head. She is relying on traditional remedies but they are not working. A shivering fever has taken grip of the whole island. So many are ill that the usual festivities marking the start of November (a month full of pageants in Panama) have been abandoned. Later, captain Toni slips Justino two ibuprofen tablets: it's the sum total of the modern medical help his grandmother is going to get.

It's no surprise so many are unwell on the Robesons. Levels of sanitation, as on all the San Blas islands, are basic. Arriving at any of the larger inhabited islands, what you see are dozens of latrines perched on the end of rickety jetties poking out in all directions. The houses face inwards, away from the sea, which is the Kuna's sole source of protein but also a giant toilet. I realized the limpid waters were not a great place to swim on my first morning when, after a few hearty strokes, I found myself trying to negotiate some serious-sized turds.

A crowd of children tumble into Justino's compound, including a girl with a shock of blonde hair and very white skin. Albinism is common among the Kuna and such children are considered lucky. Toni asks the children why they think the island keeps getting flooded. They giggle and push each other; they don't know. It turns out none of them – not even the older teenagers – has ever heard of sea-level rise or climate change or global warming.

Maybe from their point of view things don't look so straightforward. They see the islands inundated but also actively defended. As those interviews on Coetupo suggest, some of the techniques used by the Kuna to protect the islands are of dubious value. I came across an even less likely practice a couple of hours' sail south of the Robesons, on Chichime island. Here the elders had advised that sea rise could be stopped if a particular mixture of sand – with equal parts from different parts of the island – was banked up. The sand was soon washed away, and later I picked my way along Chichime's fast-eroding shore, over scattered plastic waste and the dying black roots of palm trees that were once inland.

Ill-conceived responses like this may be more the exception than the rule. Other Kuna techniques do work, at least for a while. On Tupsuit Dummat I listen to the account of Armando who is taking a rest from unloading sand that he has collected from the mainland in his *ulu*. He uses the sand to reclaim land. Even with a mechanical digger it would be a big job, but all he has is a dug-out canoe, a bucket and a spade. Armando looks weary and is frighteningly thin but the work, he says, must continue, for the land will be used for a new hut for visiting doctors, allowing them to stay on the island for longer periods, perhaps even overnight.

Turning around, I spot another man a little way out to sea, building up land. He is taking gravel and coral out of his *ulu* to make a new island, fetching up rocks from his canoe as he has been doing every day for months. This is Bernado and when he paddles back he tells me about his big plan. The island he is constructing will be used to build rented accommodation and is part of his scheme to lure back his children who, like many other Kuna, have gone to live in Panama City. Bernado's efforts are not unique: the Robesons have other tiny, self-built islets. The Kuna don't just live on islands; they make them.

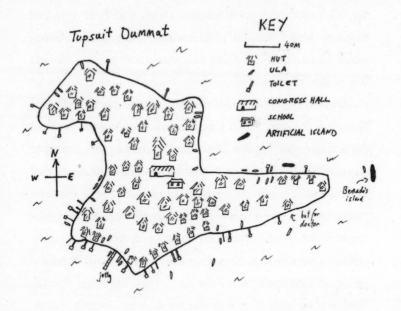

Given the doom-laden scenarios that surround San Blas, such efforts might appear futile. But they might also make us question the way these islanders' 'displacement' is talked about as inevitable, almost natural. If the Kuna had access to the resources that the Dutch are using to build Ocean Reef in Panama City, they would not have to leave. It wouldn't be complicated: these are usually calm waters, and bulking out the top and sides of these islands with rock would not be rocket science. It's not going to happen because the Kuna are poor. Not only that, they have an uneasy relationship with the rest of the country.

In order to understand the Kuna's isolation we need to delve a little into their history. In many ways the Kuna are a rare example of indigenous survival. Not only are they still here but their language and culture have flourished. Some aspects of their culture are surprisingly liberal. When men marry they move into the wife's compound and it is women who control all the money in Kuna households. It is also common to see transgender men who live and dress as women in the islands. Apparently the youngest son in a family is assigned a female role if there aren't any daughters.

Anthropologist James Howe's *A People Who Would Not Kneel* is the telling title of the most comprehensive history of the modern Kuna. An indication of what the Kuna were up against is Panama's 1912 law on 'indigenous civilization'. It opens with the proclamation that 'The Executive Power shall seek, by all possible peaceful means, the reduction to civilized life of the barbarian, semi-barbarous and savage tribes that exist in the country.' Kuna clothes, beads and nose rings were forcibly removed, while the Kuna language was repressed. The 1920s saw stand-offs and violence between the police and Kuna armed with machetes. It was a confusing time and the call for Kuna independence also reflected the activities of outsiders, most famously Richard Marsh, an American explorer sympathetic to the Kuna. Marsh wrote and circulated the 1925 'Declaration' of Kuna independence that led to the brief existence of the Republic of Tule. This would-be state was soon reunited with Panama following the mediation of the United States.

Whatever Marsh's role, the events of those years are remembered and re-enacted across Guna Yala every year and referred to as the Kuna Revolution, a proud moment of defiance and a living legacy of resistance. Rather alarmingly (though misleadingly), the flag of the 'revolution' was a bright swastika and is still a common sight in the islands. The flag has no connection with the Nazi symbol, although I was unsettled to read Eric Bauhaus's claim that the revolution was accompanied by a 'Holocausto de las Razas' in which people of mixed parentage were murdered. While James Howe acknowledges 'ugly killings of non-Indians', his account suggests a total death toll of only thirty, so the word 'holocaust' hardly seems fair. Nevertheless, a profound dislike of 'mixing' is still very much alive. I was told repeatedly by Kuna and non-Kuna alike that it is forbidden for Kuna to have relationships with non-Kuna.

Panama won the Kuna back after their revolution – as they have several times since – with promises of autonomy. 'Autonomy' sounds like a good thing and today the Kuna pretty much govern themselves. Yet, in an age of global crisis, this kind of freedom comes with a sting in the tail. Unlike other 'sinking' and fully sovereign islands such as the Solomons or Maldives, the Kuna's plight is invisible. They have no voice at any international table. Small islands like these, without independence, have a problem: few have heard of them, and their concerns mean very little to the outside world. The problem is made worse when they are inhabited by an impoverished minority in a country where environmentalism has made few inroads. In my experience, in Panama – as in so many other countries – concerns about damage to the environment are met with a shrug. The political autonomy of the Kuna makes this shrug more certain and more dismissive. At best the Panamanian attitude to the Kuna is a form of benign neglect, but it is frequently clouded with irritation – even anger – at the Kuna's stubborn refusal to join the modern world.

The Kuna's hard-won autonomy has an impact on those wanting to visit San Blas. Tourist provision is haphazard and it isn't an easy place to get to. Travellers need to show their passport at the border of Kuna territory to listless armed police; after that it's often unclear who, if anyone, is in charge. My visit coincided with a breakdown in Panama–Kuna relations. The Kuna have long been annoyed that outsiders – particularly the yacht-owners – are taking tourist dollars. Since there aren't any hotels, a lot of people stay on boats and they are all owned by non-Kuna. In Panama City I got panicky WhatsApp messages from Toni and his confederates, explaining that the Kuna Congress had made my berth on his yacht illegal. I was bewildered. Tapping my smartphone into the small hours, I ended up agreeing to a cloak-and-dagger arrangement that involved taking a taxi from my hotel at 4 a.m. but not divulging to the driver where I was going: 'If anyone ask you, say you're going to David's place, OK?' It was the strangest taxi ride of my life. I had no idea where I was going (except that I needed to meet a man called Toni), yet I was forbidden from announcing my destination. The journey was four hours long, cost $70 in cash ('Give me the money now, please') and took place in an ancient Jeep that wheezed and juddered on ever smaller and more potholed roads. Eventually, I found myself in what I was told was a port but was, in fact, a clearing by a bend on a muddy river. After a long wait in the blistering sun, I was directed into a small boat equipped with two fat outboards. It powered upriver and then into the open sea. The rain began to pour as it arrived at our destination, which I later learned was the island of Chichime. Toni emerged, padding across the sand in a bright cagoule. There followed a mystifying stand-off between him and a few Chichime Kuna, which was resolved by me handing over $75 in return for a lobster dinner. I'm pleased to say my self-pity cleared up as the rain stopped and soon I was making a cosy bunk below deck on Toni's boat.

Having read about the Kuna in the international press, I was expecting confirmation of a now-familiar narrative: that the Kuna are doughty, plucky natives who are choosing to leave their islands on their own terms. It's a nice idea but it's wishful thinking. The impression I came away with was very different, less hopeful and far more chaotic. What is going on here is slow-motion panic. The Kuna are being left to fend for themselves and to make their way through a calamity that is not of their making but that is turning their way of life upside down and inside out.

The modern, industrial world is destroying these islands. Yet they fascinate the industrial world's restless and beauty-starved citizens. Each island is unique, enchanting. It's no surprise that the San Blas islands draw in tourists and outsiders like me, searching for paradise. We circle round, always looking for another lovely thing, another photo opportunity, around and around, waving and smiling, as everything disappears from sight.

TONGATAPU AND FAFA, TONGA

'Islands come and go. Some parts grow bigger, other parts disappear. It's not humans that make it happen.' We're the only customers at the island's only bar. Tom orders another large glass of red wine, takes a thick slurp and becomes adamant that 'There are no signs of anthropogenic sea-level rise but that's all people want to hear.' I wince. He's got me in his sights. A slight and rather weary young German, Tom is goading but also confounding me: he's a marine biologist, an expert and he's self-consciously dismissive of the tide of received opinion.

Tom lives on Fafa, a *very* small square island – it's about 450 metres across – and 'off-grid' eco-resort. He studies its corals and has told me he also serves as the island's doctor, which confuses me. Fafa has just thirteen newly built, traditional-looking, coconut-thatch huts for its well-heeled visitors, each standing in a clearing of palms and tropical flowers, and it's only 6 kilometres from Tonga's capital. What would be the point of a doctor? Perhaps I've misunderstood or misheard. At the moment I'm not feeling certain of anything. Some hours later I lurch into the darkness, my bare feet snouting along warm sandy paths to my own 'native' hut.

I try to fall asleep to the shush of cradling waves. There's little wind tonight to disturb the tropical air and the only other noise is the occasional squawk of a swamp hen, querulous long-legged birds that jealously patrol the island.

A tiny island can feel like a very soothing place – womblike and a world away from danger. Yet I'm restless, gnawing over that indigestible conversation. I should have challenged Tom: what natural process could cause so much rapid sea-

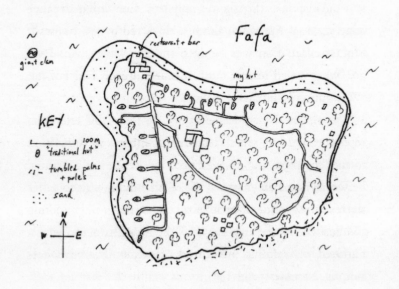

level rise? I should have said, 'Look, Tom, satellite data indicates that the sea level around Tonga has been rising every year. It's exceptional and dangerous, especially as it has been accompanied by more severe storms and cyclones.' Something like that.

On the bedside table my wristwatch is ticking. Loudly. I get up and smother it under some clothes. As the minutes drag by more rebuttals come to mind. All eighteen member states of the 2018 Pacific Islands Forum, including Tonga, have agreed that climate change is the 'single greatest threat' to the Pacific region. The plea of the Tongan government at the UN climate conference in Paris was stark: 'A lot of countries and governments are in Paris negotiating their economies – we're just asking for survival.' Just as telling was the response of the government's senior climate finance analyst, Sione Fulivai, when he was asked by journalists if a mass evacuation was being planned. 'Where would we go? We are tied to our land, to our culture,' said Fulivai. 'Without our lands who are we?'

I recall how I first heard of this place: a BBC television report from 2015, which featured an interview with Fafa's former manager and showed how the beach had retreated between 5 and 10 metres. 'We're already having to move the restaurant and bar area back,' the manager said, adding: 'We're fighting the inevitable.' The journalist asked if in a hundred years the island would still exist. The reply was simple: 'Absolutely not.'

Earlier, in bright sunshine, I'd tried to walk the island's sandy circumference and it was soon plain that this is a place under attack. Much of Fafa's southern shore is an obstacle course of palms felled by the surging waters. Attempts to stave off the rising waters are now part of the wreckage. Wooden posts, planted a few years ago to defend the shore, are now mostly at an angle or scattered and some are in the midst of water. On disappearing islands in the tropics you are often just a metre or so from storm-damaged and sea-broken trees and homes and a rapidly eroding shore – from overwhelming proof of sea-level rise and more turbulent weather.

Before I started visiting endangered islands I imagined that I'd be talking to angry people. So far, that hasn't happened. I guess if I sought out politicians or activists it would be very different. But I'm not picking or choosing; I'll talk to anyone. What I hear are sighs and shrugs – a very occasional denial that there is anything amiss – accompanied by the steady beat of waves on beaches. On the main island of Tongatapu, a guy in his sixties, who was waiting at the harbour-side, told me about the changes he'd witnessed: 'I can see the high water in the harbour; it's much higher now, so when it's high tide now it's coming over.' Then he sighed: 'I don't know if it's the sea rising or the land sinking. It doesn't matter which. Nothing I can do. Nothing I can do.'

'Nothing I can do.' We both stared out to the blue Pacific, an ocean larger than the land-mass of every single continent and every single island combined. 'Nothing I can do.'

'Nothing I can do' has begun to sound like humanity's agreed position and its solace. I'm in no position to scold or sound superior. Back in England, I recycle odds and ends, ride a bike, occasionally buy a local vegetable, things like that. It enables me to be part of the ongoing chat about being green and to think that I'm in some way 'doing my bit'. Out here, all that, it's just *nothing*. Among people who are living with the imminent disappearance of their homes – of everything – the psychology changes. It is truly overwhelming. And if any gesture captures this mood it is the shrug and the sigh, the body language of fatalism.

Perhaps I'd better check the time on my phone: 3 a.m. Still no sleep, yet I'm so tired, plagued by repetitions of half-sentences, of questions never asked or answered. Others soon bubble to the surface, such as why this glaring crisis doesn't make people jump up and down in the way other crises do. Is it because other challenges, like economic calamity or even war, are small scale by comparison and have quick fixes? This is different; it's on a different scale. Global environmental disaster is, indeed, overwhelming. It runs against some innate human feeling that nature is there for us – that everything, in the end, will be OK.

I did get to sleep; I must have as next morning I'm up early. No sign of Tom, my nemesis; I guess he won't be up for hours. The early sun has already warmed the sand and I've taken one of the resort's snorkels and am soon splashing into the water. The corals are soft and colourful, though not to be touched, and anyway I'm on the hunt for a local spectacle, the giant clam. The last time I saw one was in a comic – a monster of the deep trapping the leg of our hero. I wasn't convinced: *that* could not be real!

Through the clear shallows I make out a boulder, sand glinting on its flanks. I'm nearly over it and have become not scared but amazed: inside the long, wavy lines of its wide-open mouth is an enormous muscular siphon, pumping in and out, the black hole as wide as my fist. The clam is well over a metre long and probably getting on for a hundred years old. The world is still full of marvels.

I stumble up the beach, gleeful. I've discovered a giant! I have to remind myself that I was directed to it by one of the current managers of the resort. It is 'the clam': a survivor in waters where they used to be plentiful. Tonga still has enough giant clams to export them to aquariums across the world, but probably not for much longer. As sea temperatures rise, the shell of the clams thins and they become vulnerable to predators. Across Asia and the Pacific their numbers are in free-fall.

Before leaving Fafa I talk some more to the resort's managers, a married couple who are more guarded than their predecessor about the future of the island. A super-friendly and efficient Australian husband-and-wife team who have decades of experience managing island resorts, they point out (like Tom) that the island is shifting, building up in some places and eroding in others. Reassurance – we all want it: everything will be OK.

Fafa is an enchanting place. The mood is different on what Tongans call the 'mainland', the island of Tongatapu. It is home to 70 per cent of Tonga's 108,000 people. The remaining 30 per cent are scattered across thirty-five of the archipelago's 169 islands. Unlike multiracial Pacific nations like Hawaii, Fiji or Tahiti, Tongans are nearly all Polynesian and take considerable pride in never having been colonized. Tongatapu is 40 kilometres long and 20 kilometres wide; at its highest point, near the airport in the south-east, it rises to 65 metres. It is mostly much lower, especially on its northern shores where the capital, Nuku'alofa, sprawls along the seashore.

I pick up a hire car in Nuku'alofa and set off on a day-long trip round the island. The road I'm on is often banked up above the surrounding fields and it pretty much circles Tongatapu. There are sporadic sea defences: bits of the north coast have barrier walls or are lined with great white boulders. But it's the upgrading of this road – from a dirt track into a tarmac-covered and rigid earthwork – that has (albeit unintentionally) created the island's most significant sea-wall. It's not an unmixed blessing: the road is making storm floods worse as it blocks rainfall from draining into the sea.

The road is getting emptier as I head east, looping round Fanga'uta Lagoon; a marshy body of water that scoops out the middle of Tongatapu and provides an access route for the rising and stormy ocean to penetrate deep inland. My first stop is to see the fishing pigs of Talafoʻou. On the shoreline I pull up near a tourist information board that describes how local pigs love swimming and snuffling for seafood. But the pigs are nowhere to be seen. Maybe the tide is too high. Since they 'fish' as well as swim, they sound more worthy of fame than the more celebrated but lazier swimming pigs of the Bahamas and Bermuda.

A family in a pick-up – the vehicle of choice for many Tongans – slows down to stare at me. I get the feeling that tourists have become a rarer sight round here than swimming pigs. The absence of tourism was obvious almost as soon as I arrived in Tonga. The taxi driver who met me at the airport – a man vast in all directions – began with the question 'You an aid worker, yeah?'

The flight got in at midnight and the drive to the capital proved a sombre introduction to the country. We had to make several detours as police curfews were in force and many roads blocked off. After the last cyclone a night-time lockdown round the centre of town was thought necessary to prevent looting by what the taxi driver called 'bad boys'.

Tonga's problems are interlinked. Rising sea levels, more frequent and more damaging cyclones, a rise in crime and social discontent, large-scale emigration, a collapsing tourist industry ... they are an interlacing burden of worry. We can add other problems to this weighty chain, such as an agricultural crisis, the increasing need to import food, and an obesity epidemic. Worsening and more frequent storms batter farms, blowing away crops and killing the soil with salt water. The result is that very little fresh green produce is grown in Tonga. Today Tongans have the unfortunate international status – along with other Pacific nations who face similar challenges, such as the Cook Islands and Nauru – of being one of the world's fattest people. When you look around the food shops, which are full of packets and cans of imported food and none of which sells anything fresh, you can see why.

I drive down the east coast, passing through tranquil coconut groves, extraordinary ancient graveyards, a 'stone henge' called Ha'amonga 'a Maui, and long empty beaches. Tonga is a very old and special place, with a history reaching back some three thousand years. Before long I'm on the southern coast, which is rockier and safer from the impacts of cyclones and sea-level rise. If more islanders lived on this side, rather than on the shallow north side, lives and livelihoods would be saved. The rocks of the southern coast are perforated with hundreds of blowholes. I find a vantage point where, looking both east and west and for miles up and down the coast, you can see, hear and feel powerful white jets of sea water being gunned high up into the air.

Elsewhere the rocky nature of the beaches has another cause: all the sand has been stripped off for use in the construction industry. A further casualty of development has been the mangrove swamps that once protected Tonga's coasts from storms and tsunamis. The many problems facing Tonga have been made more intractable by the fact that all the land is owned by the Crown. In 2010 Tonga stopped being an absolute monarchy and is now described as a constitutional monarchy. However, King Tupou VI and the noble families that constitute his court still literally own the country and that has made it difficult to relocate farms and people away from the vulnerable, low-lying northern coast. Writing about the need for relocation inland, the climate-change scientists Patrick Nunn and Nobuo Mimura warned, over two decades ago, that 'If the King and his nobles are not prepared to release more land to commoners for settlement, trouble may arise.'

Trouble has arisen but not against the monarchy. Discontent has been displaced onto the great bogeyman of twenty-first-century Tonga: the Chinese. The rioters who destroyed about 70 per cent of the capital's central business district in 2006 targeted Chinese businesses. Resentment against Chinese loans and business ownership continues to seethe. If you go into any store in Tonga, you're likely to see a young Chinese woman staffing the till, while glum Tongan employees and their friends hang out some distance away. You can almost taste the hostility in the shop-worn air.

It was only in October 2018 that Tonga got a sea-level monitoring station, and detailed information on the scale of the problem remains in short supply. Most people appear to agree that flooding from sea level is a less pressing issue than cyclones and the threat of earthquakes and tsunamis. Every corner I turn brings fresh evidence of the destruction wrought by the last cyclone. A lot of Tongans live in houses of breeze-blocks and corrugated iron, which keep getting blown down. Some homeowners have slung plastic sheeting up to keep off the rain but others have moved out. The risk of flooding, tsunamis and earthquakes in Tonga is extreme. Not only does Tonga sit in one of the most cyclone-prone areas of the ocean but it is also in the middle of the 'Tonga Trench', the world's most active zone of tectonic movement. This is a landscape well used to disaster. The island of Tongatapu periodically is buckled and tilted by earthquakes, such as the one on Christmas Eve 1853 that caused the whole northern coast to subside and flood. If you want to see what tsunamis can do to the island, a visit to Tsunami Rock is instructive. It's a boulder the size of a large house. Covered in small trees, it squats incongruously in the middle of a field. It was thrown up, far from the coast, a thousand years ago by a massive wave that swept over the entire island.

Earthquake and tsunami drills and planning are a regular feature of life in Tongatapu. Smartly uniformed schoolchildren have afternoons set aside in which they practise running to what passes for high ground. The Tonga Meteorological Service has a network of seismic stations but relies on the Pacific Tsunami Warning Center in Hawaii to

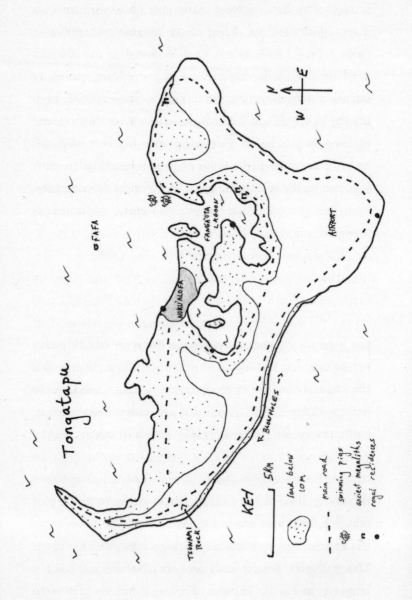

provide information that is early enough to save anyone. Tongans prepare for climatic and natural calamity on a daily basis. They have to. The problem is that they face a whole slew of interconnected issues – from salt-water intrusion to earthquakes, from crime to sea-level rise, from obesity to ethnic conflict. All thrown at a very remote and poor country of barely 100,000 people. In many ways, it's remarkable how well Tonga has coped. It hasn't fallen apart – not quite; it's still there. But the challenges are immense. The three-thousand-year history of Tongan civilization is slowly unspinning.

THE ISLES OF SCILLY, UK

It's a peculiarly warm and cloudless day in late February and I'm trying to take in all the rocky islands that freckle the small bay and the far horizon. My daughter and I are the only guests in a seafront guesthouse with a little balcony. Steve, the owner, has some helpful words of explanation.

'What you're looking at are the tops of mountains,' he tells us. For thousands of years the Isles of Scilly have been drowning. It's a flooded landscape, a vision from the past that is a fingerpost to the future. Leaving the balcony, we head to the holiday flat's kitchen, which has views over the Lower Moors: rough fields and reed beds that occupy a large wet chunk of the main island of St Mary's. There is a more urgent note in our host's voice: 'I'm hoping they will be there for a good while; that's the hope.' But later that day he produces a battered copy of the local council's 'Climate Change Strategy' plan, which includes a map in which the whole of the Lower Moors – and much else besides – is coloured in blue, given up to the Atlantic in what the document calls 'managed retreat'.

Since that plan was published in 2011, it has become clear that allowing the island's only freshwater bore-holes, which are on the Lower and Higher Moors of St Mary's, to be swallowed up by the sea might not be such a great idea. It's possible another fifty years could be bought for the Lower Moors with coastal 'rock armour' that will help reduce flooding during storm surges. For the time being at least, there is a determination to stay. This is, after all, a landscape that has been living with loss for a very long time.

The Isles of Scilly is an archipelago of 200 or so islets and islands, 43 kilometres west of England's western tip, and when the weather is kind – like today – you can understand why some call the Scillies the 'Fortunate Isles'. Its land-mass is tiny, just 16 square kilometres, but it feels less crowded than the mainland. The population of 2204 is small and stable. In fact, there are fewer residents today than in the early nineteenth century, when the population reached about 2500. There are five inhabited islands (St Mary's, St Martin's, Tresco, St Agnes and Bryher). Even the biggest, St Mary's, is easily walked round in a day. Not that you would want to rush: the entire island group has been designated an Area of Outstanding Natural Beauty. Many of its quiet roadside verges drip with succulents and shy pretty flowers and the Scillies has by far the greatest density of historic sites in the UK, with 239 ancient monuments and archaeological sites on land and many more lying uncounted under the waves.

Before the Scillies there was Ennor, Cornish for 'The Land'. Ennor was a big island whose higher peaks are now the main islands of Scilly. Ennor was always shrinking. St Agnes was a separate island by 3000 BC and it is estimated that in the 500 years between 2500 and 2000 BC Ennor lost over half of its area. The name 'Ennor' was still in use in the sixteenth century, and everywhere you look on the Scillies you can see its phantom traces. Ancient stone walls run into the sea; multiple tombs cluster on rock islets that used to be hill-tops; the ruins of the Iron Age village of Halangy sit on a low cliff, looking out over a seascape that would once have been farmed. From 1500 BC onwards most of this low-lying farmland turned into marsh. Charles Johns, from Cornwall's Council Archaeological Unit, explains that it would nevertheless 'have remained useful land, especially for grazing animal stock and would have been passable with ease almost all of the time'.

It took a long time for Ennor to completely fade from view. Charles Thomas, whose book *Exploration of a Drowned Landscape* is the definitive guide to the island's deep past, tells us that the final transition from Ennor to the Scillies 'took place between the seventh and thirteenth centuries, and that "from the viewpoint of people in small boats" the waters between the present islands would not have been navigable until Tudor times'.

Sea-level rise in the Scillies is a compound of two long-term processes and one recent one. The two long-term processes are an interglacial, warming period and post-glacial 'bounce-back' further north, which is tilting the southern half of Britain downwards (this 'isostatic adjustment' adds between 10 and 33 per cent on existing sea-level rise). The short-term process is modern, anthropogenic global warming, which is making a challenging problem not just worse but unpredictable.

It's not incremental rises that the islanders fear so much as storms and sea-surges, which regularly flood the island's capital (and only) town, Hugh Town, which straddles a sandy isthmus on St Mary's. The storms fling up salt water and boulders and eat away at coastlines, reducing the viability of the island's petite daffodil fields and threatening the tourist economy. The litany of recorded storms – from the Great Storm of 1744, through to the storms of 1962, 1989, 1994, 1995, 2004 and 2014 – suggests they may be becoming more regular. The last big one, in February 2014, ended up with the main street of Hugh Town covered in sand.

I've come to the Town Hall to talk to Julian Pearce, the council's Physical Assets and Natural Resources officer. The Town Hall is in the centre of Hugh Town and the hub of island life. There are posters inside and out, telling islanders to conserve water. The message could not be clearer: 'Water usage has reached an unsustainable level on St Mary's.' I have, though, a more lightweight message to deliver: Steve has told me to remind Julian about their badminton practice. They are both spry, middle-aged men so I can imagine it will be quite a workout. It's only later that I realize that my errand might be a clue to something important. It's no exaggeration to say that everyone here knows almost everyone else. 'Community' is an overworked word but in small islands it means something. The fact that there is a close-knit community explains why the Scillies, and other small islands, battle on. Julian tells me how 'the community still gets engaged', which in practical terms means that when there is just a few hours to prepare before an incoming surge tide, people get moving and start shifting sandbags.

To drive the point home he tells me a story about a meeting of a would-be flood-preparedness group that was called in Newquay in Cornwall, on the mainland. The assembled citizens were informed that seafront properties were in real danger of being swept away. 'It was supposed to be about getting the neighbourhood together,' says Julian, but 'the people living further back said: "So those houses are going to go? Fantastic! We can have seafront properties!"' Julian is clearly relishing the tale. Islanders know they are different and they define themselves against 'mainland' selfishness. Perhaps this also helps explain why, despite the dangers of sea-level rise, people want to move to islands and build islands: islands symbolize real community and the kind of values that so easily dissolve in the acidic anomie of mainland life.

There are some big decisions looming on the Scillies and they will test its sense of community. Julian is cagey about Hugh Town's longevity but he is clear that 'in x hundred years' it will be covered by sea-level rise. Yet 'managed retreat' is no longer a favoured phrase. Referring to the map with the big blue areas of 'retreat' that Steve showed me earlier, he cautions that 'that approach was based on defending property' and says today they need 'a better and more balanced viewpoint' that also considers essential infrastructure. Julian points out that the 2011 plan 'didn't look at water supply or access to airport or communication from town to up-country'. ('Up-country' is what people on St Mary's call the part of their island that isn't 'town', which is Hugh Town.)

Identifying problems with 'managed retreat' is one thing; coming up with an alternative is more tricky. Walling in the Scillies is not an option and, citing plunging sea depths round the archipelago, Julian is not keen on barrier islands. If the funding comes through, he is pinning his hopes on rock armour in parts of the north-west of St Mary's: 'That would give us another fifty years, say.'

I cast my mind back to the *Daily Telegraph* storyline that first attracted my attention to the Scillies: 'Scilly Isles could have to be abandoned because of global warming'. I also think back to the Skype call I had with Jan Petzold, who is a Science Officer for the Intergovernmental Panel on Climate Change, based in Bremen, and an expert on the Scillies. Petzold is concerned that all the attention on Pacific nations is giving people a false picture of climate change. 'We don't know much about islands in the north,' he says, even though 'we have a lot of islands in Europe and along the American coasts and people there are also vulnerable'.

The next day it's still hot enough for early summer (it will be confirmed later that this is the warmest February since records began) and I take the path to Old Town, a village that is overlooked by the ancient Ennor Castle, in order to meet up with Nikki and Darren from the Isles of Scilly Wildlife Trust. The Trust looks after 60 per cent of the Scillies. They have a ninety-nine-year lease from the ultimate power and landowner in these parts, the Duchy of Cornwall, the yearly payment for which is one daffodil. Under the Trust's care are all the uninhabited islands, the coasts of the inhabited ones (except for Tresco, which is leased to the Dorrien-Smith family), and St Mary's Lower and Higher Moors. Nikki and Darren are interested to know what Julian has told me and what his plan is. I am a little flustered to be imparting potentially important information – messages about badminton practice are more my style – and garble something about the errors of 'managed retreat', to which they readily agree. It seems I've come at a key time: old plans are being folded away but new ones have yet to be drawn up. Another complicating factor is that the Isles of Scilly Council is in the process of handing over responsibility for fresh water and sewage to a big supplier, South West Water. South West Water are promising to invest £40 million by 2030 but I doubt that protecting the islands from sea-level rise is part of their remit.

We walk beside a ditch that divides the Lower Moor. On one side are brackish water and an endemic species of sea rush; on the other are fresh water and common reed. Darren, the Trust's Head Ranger, is clad in waders; he's just come from chainsawing willows that are threatening to block the ditch. The Moor is hardly noticed by tourists keen to rush off to other islands but, for Darren and Nikki, this is a key landscape. We pass a concrete bunker where a fresh-water bore-hole draws up its precious resource. Keeping the Moor as wetland by controlling the water-level and managing its vegetation is essential not just for the human population but for many other species too. We hunch down on a walkway made of recycled plastic, and Darren draws up pond species found in few other places in the UK, such as Tubular Water-dropwort. Overhead he points out the black, gliding shape of a glossy ibis, a Mediterranean visitor which is becoming a more common sight as the climate warms. I can't help thinking that this is a Wildlife Trust like no other. It is working for both the human and non-human population and sees them as intimately connected.

During the tourist season the seas around St Mary's are full of chugging boats, ferrying visitors hither and thither. As soon as people land on one island they start thinking about another. Not many boats run in February and, anyway, I'm keen to get to Halangy and Bar Point, the northern sprig of St Mary's, which old maps show as offering a crossing point, a drowned road.

Julian tells me that, when his daughters were small, they would play at Halangy, calling it their fairy village. I'm here with my daughter, Aphra, who has moved on from her initial cynicism at the idea of my island adventures. I think she was hoping to join me on a more exotic destination but the Scillies do feel a world away, especially in this freakishly hot February. Aphra's too old to play fairies – in fact, I don't think she has *ever* played at fairies – but she's having fun leaping between the stones of the Iron Age village of Halangy. We drop down to the beach where signs of other ancient settlements have been found in the cliff face and have for many centuries been pulled apart and washed away by the rising sea. When we get to Bar Point there is plenty more evidence of prehistoric occupation, including a number of Bronze Age burial cairns. Aphra climbs on top of one that sits right on the coast. When built it would have been the summit of a hill surrounded by a broad valley laced with walled fields. The Scillies are so full of ancient graves that it has been speculated that important dead people from the mainland were rowed out here for burial. The landscape has so many layers of loss that, in quiet moments by the shore, it can feel a little odd that a tourist economy promising to deliver 'happy memories' has now been rolled out, smothering them all.

Back on the balcony of our holiday flat, my eye is drawn to the empty island of Samson. It has two low hills and is not inhabited nor much visited. For many thousands of years, however, it was home to untold generations, as testified by the many ancient sites and field walls scattered on it and around it in the sea. From the sixteenth century it was occupied and farmed by two large families, the Webbers and Woodcocks. The population was between thirty and forty by the end of the eighteenth century, though drought was common, meaning the island's small aquifer had to be supplemented by water brought from neighbouring islands. A Baptist minister called Bo'sun Smith (also known as the 'Seafarer's Apostle') visited in June 1818 and found a sorry scene. 'Two or three families are very poor, and have suffered much distress,' he reported, adding: 'their chief support has been limpets, as the immense piles of empty shells before their doors sufficiently testify.' The population was described as 'poor but industrious and pious'. So pious, in fact, that they may have been the last people in Britain to retain the Julian Calendar – celebrating 'Old Christmas' on 6 January – up until the time the island was finally abandoned in 1855. The islanders were given no choice but to depart, coerced into leaving by Augustus Smith, the Scillies' Lord Proprietor. Last to go was Mrs Webber who, it was said,

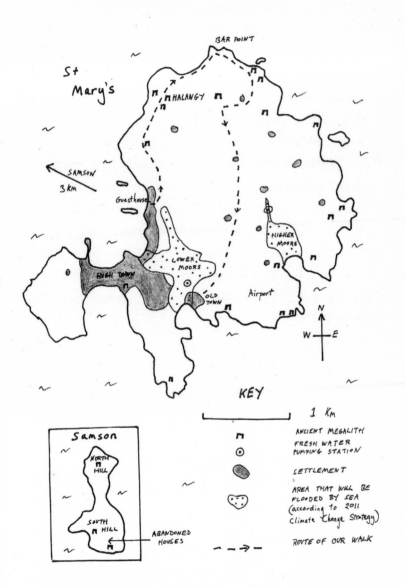

St Mary's

BAR POINT

HALANGY

SAMSON
3 KM

Guesthouse

LOWER MOORS

HIGHER MOORS

HUGH TOWN

OLD TOWN

Airport

N
W — E

KEY

1 KM

⌐ ANCIENT MEGALITH

◉ FRESH WATER PUMPING STATION

⬤ SETTLEMENT

AREA THAT WILL BE FLOODED BY SEA (according to 2011 Climate Change Strategy)

– – → – ROUTE OF OUR WALK

Samson

NORTH HILL

SOUTH HILL

ABANDONED HOUSES

knew magic. As Charles Thomas retells the story, she 'put a spell on' Augustus Smith: 'His legs would not move, he was unable to get back in his boat, and the Tresco boatmen had to persuade the lady to lift the enchantment.' For Thomas, 'among all those strange nuances of atmosphere that Samson still breathes, grief is perhaps the strongest'.

Over its long history, the Isles of Scilly must have seen many abandonments. If it doesn't want more then 'managed retreat' will have to give way to resilience and protection. Easy to say. Especially for fair-weather friends. In a few days I'm scuttling away on a fifteen-minute plane ride, back to Land's End airport. I'm travelling away from a pretty place but I'm not leaving paradise. Like so many islands, fragility and melancholy is threaded through the beauty of the Isles of Scilly.

PART THREE

FUTURE

Future Islands

WHAT ARTIFICIAL ISLANDS will be built over the next fifty years? In order to understand where our age of islands is heading, I want to look at plans for three very different artificial islands: a free-trade utopia (Seasteading); a 6-square-kilometre energy hub (Dogger Bank Power Link Island) and one designed to give homes to over a million people (East Lantau Metropolis).

Will the appetite for new islands ever be sated? It seems we're not going to stop building them anytime soon. The end of the conveyor belt that is dropping islands off the coasts of the Gulf States and China is still not in sight, and many other countries – especially newly aspiring ones that want to attract and keep wealth close to their shores – are keen to get in on the action. They no longer look to places like Britain or the USA as models of development but to Dubai and China, and new islands are part of the package. One of the most ambitious is Lagos's Eko Atlantic, a 10-square-kilometre island of high-end apartments and stores. Designed for 250,000 residents, with a boulevard based on the Champs-Élysées, Eko Atlantic offers a haven for Nigeria's wealthy. They will be safe from the sea (the island is defended by the 8.5-kilometre 'Great Wall of Lagos' flood defence) and from the envious eyes of the rest of Lagos. These are islands of affluence; more specifically, they are examples of what some geographers call 'secessionary affluence': leaving the problems and the people of the ordinary city far behind.

Spin the globe to newly affluent economies further east and we find plenty more islands in the pipeline. One of the most spectacular will be Forest City, set in the Johor Strait between Malaysia and Singapore. It began in 2014 and its four artificial islands are designed to accommodate 700,000 people. As I write this, the first island, Country Garden Island, is already visible on Google Earth as a rather bleak building site, with a striking green patch where turf and tree-lined walkways have been rolled out around a hotel. The sales pitch for Forest City has an ethnic target. In one of the promotional videos a Chinese couple tell us that 'Many ethnic Chinese live here. For us, it's more like we're living in our hometown than a foreign country.' The idea seems to be that well-heeled Chinese, who are fed up with the pollution and crowds of Chinese cities, will want to invest in this cleaner, greener alternative. It almost goes without saying that another striking feature of Forest City is how much forest and other natural habitats have been destroyed to make it. Instead of mangrove forests there are golf courses; instead of seagrass meadows, there are planted road verges and gardens.

The offshoring of wealth and leisure demands both the presence of nature and that nature be ripped apart. These upmarket islands are outposts of a predatory and controlling infatuation: we love nature so much we have to own it, kill it then reproduce it in forms so small, so reduced, that it is graspable. The plans and pictures that sell these planned islands make them look like green fantasy lands. The painterly images produced for Toronto's new Villiers Island, which has the Don River winding through its middle and navigable canals on either side, show acres of flowering meadows and joggers joyously bouncing through canyons of trees. Property agents gush that 'Villiers Island will be roughly 88 acres in size and will be an ecological masterpiece.' However, this is a development that will provide the city with flood defence, as well as new land for houses and parks. At this stage, it does seem genuine in its attempt to strike a balance between environment, infrastructure and housing. It is being built around an urban river and on abandoned city lots. Unlike so many new islands, this is not a pristine habitat about to be wrecked but somewhere that a new island could have real environmental value and serve the city.

Today, environmental damage is one of the key reasons that island-building plans are likely to founder. Island-building projects are big, complex endeavours and there is ample opportunity and time for one partner or another to stall things or pull out. In the late 2010s the island-building market was showing signs of nerves. Plans were shelved, then dusted off again, only to be shelved once more. One plan that caught my eye was in Slovenia. Slovenia has no coastal islands: a point rubbed in by the fact that its southern neighbour, Croatia, has built a tourist economy on the back of having more than 1200. When it comes to tourism, countries with lots of islands are regarded as blessed. It follows that countries without any might want to build one. Slovenia's plan was modest: an island about the size of a large shopping mall, offering beaches, bars, restaurants, a wellness centre and a marina. But the finance for the new island comes and goes and politicians are uncertain. The project is in danger of turning from a grand hope into a national embarrassment. The same can be said, on a much bigger scale, about the Federation Islands, a 330-hectare archipelago in the shape of the Russian Federation planned for the Black Sea. After much fanfare, this great patriotic project stalled in 2012 and seems to have been officially forgotten. Jakarta's Great Garuda, a seventeen-island sea defence built in the shape of a giant eagle that was to be home to 300,000 people, has gone the same way. Once a centrepiece of national pride and not a little hope, its sea-wall – at 40 kilometres long and 24 metres tall – was supposed to bring long-term security and new housing land to the rapidly sinking city. But politicians began to look at it not as an eagle but a white elephant, and in 2018 the funding was pulled.

There are plenty of reasons to get cold feet. It often occurs when someone dares to ask what this shiny new island will look like in fifty or a hundred years' time. The blueprint that developers and politicians were enthusing over a moment ago suddenly seems less attractive. Some bright spark might venture that one solution to sea-level rise is to build floating platforms: they are cheap to construct and highly adaptable. But floating islands are dependent upon links to the mainland. They are also very vulnerable to storms and have a short shelf-life. The leading authority on floating platforms, Professor Wang Chien Ming, points out that platforms can only last fifty to a hundred years and that 'nobody would want to live there after a hundred years'.

In democratic societies, plans for new islands are always the subject of intense debate. Public controversy derailed plans for Belgium's first artificial island in 2018. Its supporters said it would protect the coast; its detractors said it would turn local beaches into rubbish-strewn dead zones. The mayor of the town next to the proposed island said: 'I was in Dubai to see how they handled it there. I passed a sewer full of plastic and oil. I wouldn't want to spend my holiday there.'

Because the role models come from the Gulf States and China, artificial islands come weighed down with certain images: of heedless consumerism, authoritarian government and environmental irresponsibility. In the West, the fame of these high-profile examples is acting as a brake on development. Since islands can protect coasts and create new land not just for people but for trees and wildlife, blanket hostility to them is short-sighted. The proponents of Belgium's island have thrown the question back to the objectors: if not this, then what?

As we have seen, most artificial islands cater for short-term and short-sighted human ambitions. As long as a sea view continues to attract premium property prices, the strange, paradoxical nature of our age of islands will persist. The need for infrastructure islands for dirty and noisy industries is also unlikely to abate, nor the military and territorial value of artificial islands, which has been proved in recent years in the South China Sea. Our age of islands is thrumming with activity: plenty are being built and plenty more are in the pipeline. But whatever they are used for, twenty-first-century islands will require a lot of maintenance: pumping stations will need to be oiled and ready to work round the clock, and sea-walls will need to be high.

SEASTEADING

Plans for the floating, libertarian city of Seasteading are well advanced. With the help of wealthy backers it is good to go; if only it could find a home. For a while it was envisaged Seasteading would land off the coast of

Honduras, then a lagoon in French Polynesia, where high-level conferences and a 'memorandum of understanding' seemed to have cemented the deal, with building to start in 2020. But protesters objected to the environmental impact of their would-be new neighbour, accusing Seasteading of being an elite plaything. In February 2018 the government of French Polynesia declared that the deal was off.

Seasteading is a tenacious project driven by the conviction that a free, mobile and entrepreneurial (or, as Seasteaders say, 'aquapreneurial') utopia is just within reach. Founded in 2008 with the help of a $500,000 donation from the co-founder of PayPal, Peter Thiel, it proposes a new model of citizenship based on 'dynamic geography'. Flexibility and choice are the guiding principles. The Seastead is conceived, as far as possible, to be separable into individual units that can relocate at will. The initial plans envisaged '50-meter-sided square and pentagon platforms with three-story buildings'. The platforms would be made of concrete and 'could be constructed for approximately $500/square foot of usable space'.

Concrete floating platforms, usually held in position by anchored wires, have been in use for thirty years and are a proven technology. The innovation was not in the engineering but in the overall ambition. Seasteading is a way of life, a movement, that wants to reimagine our relationship to government and territory. By preference, its destination is the open sea, since in international waters Seasteaders would not be thwarted by government. Beyond territorial waters, a new beginning can be made. Seasteading was founded by Patri Friedman, the grandson of the famous, Nobel-prize winning advocate of the free market, the economist Milton Friedman. Patri explains that 'When seasteading becomes a viable alternative, switching from one government to another would be a matter of sailing to the other without even leaving your house.' In the project's principal research report, it is explained that 'If modular ocean homes and offices are mobile and can be reassembled according to individual preferences, small groups of entrepreneurs and investors can feasibly build "startup" societies on earth's last unclaimed frontier.'

Although 'movability all the way down to the size of a single autonomous house' is the ideal, the authors of the project's report are aware of the problems that multiple connections between lots of individual units would create and so they propose a variety of configurations. The Seasteaders are not wedded to any one design: 'Seasteading' is a verb rather than a thing, it is a practice that can be adopted and adapted, allowing 'the evolution of new societies and forms of governance'.

Seasteading: How Floating Nations Will Restore the Environment, Enrich the Poor, Cure the Sick, and Liberate Humanity from Politicians is the ambitious title of Seasteading spokesperson Joe Quirk's book-length manifesto. It's worth drilling down into that long subtitle.

"Scenarios of Relocation"

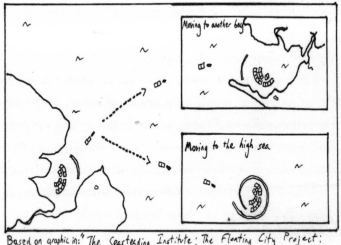

Based on graphic in: "The Seasteading Institute: The Floating City Project: Research Conducted between March 2013 and March 2014"

'Cure the sick'? In a series of web shorts, Quirk explains that 'our parents' regulations are preventing today's innovations, and that today: 'In the United States it takes ten years and a billion dollars to bring a new drug to market'. Free of all that red tape, seasteads will be places where new medicines can be created. 'Enrich the poor'? Seasteading takes inspiration from small island nations that have been rapidly transformed from poor to rich. To a backdrop of glitzy photographs of Singapore and Hong Kong, Quirk's message is that 'every time a new island nation hits restart with new rules based on modern knowledge, the poor create their own wealth'. 'Restore the environment'? This claim is based on a variety of green-tech solutions, most notably the use of the temperature difference between deep ocean water and warm tropical surface waters to produce electricity (ocean thermal energy conversion) and the farming of algae-based biofuels.

There is a bubbling energy to the seasteading movement. I'm not attracted to its ultra-mobile, aquapreneurial, political programme. But then I'm not much of a libertarian: from what I've seen, when we turn our back on government the outcome is an even worse kind of tyranny. Yet I can't help admiring Seasteading: like all the best utopias, it combines big ideas with an obsession with the details.

A willingness to adapt the grand plan in order to get the ball rolling led to the 'memorandum of understanding' with French Polynesia and an agreed site, a sheltered lagoon in French Polynesian waters. The French Polynesians were interested in hosting an innovative 'smart city' that could bring in jobs and money. For their part, the Seasteaders have come to accept that international waters are a tough place to launch their dreams. Big waves and deep waters mean that small-scale floating structures would be tossed about like a cork. A larger city would fare better but as a first step that was never likely and, in any case, there is still much to learn about how any such structure would cope with rough water. A Seasteading feasibility report notes that 'semisubmersibles and breakwaters are the options most suitable for the open ocean' but that 'the costs of a breakwater are prohibitively expensive'. The idea that individual units would be detachable and be able to sail away at any time, adds on even more costs. The favoured plan for the Polynesian site was a decidedly fixed-looking horseshoe of individual floating homes joined by short pontoons to a large multi-storeyed platform with shops and offices. The promotional film made to sell the project doesn't even mention the detachability of units. The emphasis is on convincing the outside world that this is an environmentally sensitive proposal and that it would be self-funded.

So what next for Seasteading? Its relationship to the Pacific has been kept alive by Marc Collins, a Tahiti-based entrepreneur who was instrumental in first getting the Seasteaders to locate their operations to French Polynesia. Collins's company, Blue Frontiers, caused a lot of media interest when its plans for a sustainable floating city were unveiled at the United Nations in New York in 2019. Blue Frontier's focus is on creating self-sufficient, environmentally friendly cities that can survive an era of climate chaos. The libertarian politics is no longer quite so prominent. But Collins is very much a seasteader: it's an idea, an identity, that is proving adaptable and tenacious. A seastead city is still probably a long way off. But this movement has already, in part, succeeded: the idea is out there.

DOGGER BANK POWER LINK ISLAND, NORTH SEA

The North Sea is rough and cold but it boasts the world's largest collection of wind farms. All these turbines need regular maintenance and long cables connecting them, so it's no surprise that a new 'Power Link' island is being planned at the heart of the North Sea, about 100 kilometres off the east coast of England. The proposed island is perfectly round, with a key-shaped harbour running into its centre to allow sheltered anchorage. It will also have workshops, accommodation blocks and a runway. Dogger Bank Power Link Island is designed to be a place staff can live on rather than just endure, so it also has an artificial lake and a cluster of trees and green spaces. The planned island is 6 square kilometres: a little smaller than Gibraltar. It is a major undertaking and, once completed, will transform how we think about the North Sea. At the time of writing, work on the island has not started but the plans are well advanced and backed by serious money. It is estimated it will take seven years to build and is scheduled for completion between 2030 and 2050.

Over 70 per cent of Europe's offshore wind installations are in the North Sea. Rob van der Hage, who works for one of the energy firms investing in the project, explained to the *Guardian*, 'The big challenge we are facing towards 2030 and 2050 is onshore wind is hampered by local opposition and nearshore is nearly full. It's logical we are looking at areas further offshore.' Going that far out to sea also means that turbines can take advantage of the fact that wind speeds are generally faster and more constant in open water. Distance from shore normally means greater water depths, but the North Sea has an unusual under-water topographical feature: a range of underwater sandy hills, called Dogger Bank. In this shallow zone the water is just 15–20 metres deep, which makes it far cheaper and safer to build on.

North Sea Wind Power Hub

Artificial lake

"Power to Gas"

Helipad

grasslands

wood

Sand

harbour

rock armour

based on film released by North Sea Power Hub "North Sea Power Hub vision" northseawindpowerhub.eu

Dogger Bank Power Link Island will sit at the centre of 10,000 wind turbines. It is the brainchild of North Sea Wind Power Hub, a consortium of European power suppliers (TenneT and Gasunie in the Netherlands and Germany and Energinet in Denmark), with additional collaboration from the Port of Rotterdam. Torben Glar Nielsen, Energinet's technical director, told the *Independent*: 'Maybe it sounds a bit crazy and science fiction-like but an island on Dogger Bank could make the wind power of the future a lot cheaper and more effective.' The word 'crazy' was picked up in the newspaper's headline but it's an increasingly old-fashioned perspective. Artificial islands have long since ceased to be outlandish. The Dutch are well used to building islands and, given the growth of turbines in the North Sea, the proposal is long overdue.

The island will be built at the far end of Dutch waters, so it will be within the Dutch Exclusive Economic Zone. If it was moved a bit to the west it would be in British waters, but this is a Dutch-German-Danish initiative rather than a British one and it is Europe-facing. It seeks to create an integrated cross-border network of sustainable energy supply. It's interesting to note that the island would give the Dutch a case to extend their territorial waters westwards into UK waters (since, once built, Dutch waters would extend from the island). This is not a topic that has even been raised, at least not publicly, but it's a reminder of the importance of being involved early on in big schemes on your doorstep. If ever relations between the UK and the rest of Europe were to get even worse than they are at present, and the lines of ownership in the North Sea were to come into dispute, Dogger Bank Power Link Island would quickly come to be seen as more than just an energy hub.

Looking at the plans of the company that has been doing most of the legwork, the Dutch state-owned TenneT, it's clear that Dogger Bank Power Link Island is a long-term destination and that there will be other, smaller hub islands built before it. These will also be in Dutch waters but closer to home, feeding energy back into what the company identifies as European 'load pockets'. Load pockets are zones of high demand, and the companies involved in the Dogger Bank project have identified six: high-demand, densely populated areas of England, the Netherlands, Germany and Belgium, with the North Sea Hub sitting at their centre.

The bottom line for the Dogger Bank island and its smaller forebears is that they will reduce costs. Running off-shore wind farms from the mainland is expensive: constant ferrying of people and equipment over choppy waters adds up and so does having to stretch submarine power cables over such great distances. Having the island on Dogger Bank will shorten all sorts of connections and make it possible to send off energy in different directions, enabling the island's owners to trade power in multiple markets. The involvement of the gas company Gasunie, which is also owned by the

Dutch government, tells us that the island will not just be for electricity; the idea is that it will also enable 'power to gas'. What this means is that some of the electricity generated by wind power will be turned into gas. This can be done by electrolysis: splitting water into hydrogen and oxygen. The hydrogen can then be stored underground in empty oil and gas fields. The point, as always, is cost. Gas is much cheaper to send and store than electricity.

Tens of millions of people pass through infrastructure islands, such as airports, but most are out of sight and out of mind. We are increasingly intolerant of seeing, smelling and hearing the consequences of our industrialized lives – the garbage depots and chemical works we all rely on – and the twenty-first century is likely to see many more planners reach for the neat but expensive solution of offshoring. A lot of people in the crowded nations of north-west Europe don't want huge wind turbines marching over their precious countryside. But that's not the only thing going on here. The North Sea is being reimagined: turned from an empty in-between space into a heartland. It sits at the centre of one of the world's most populous and wealthiest regions. In every one of those nations people are scratching their heads, trying to find new sources of clean energy. Reimagining the North Sea as no longer a zone to hurry over but as a hub – a site of connections and networks, sending out power in all directions – is ambitious, even visionary, but it's also sensible.

EAST LANTAU METROPOLIS, HONG KONG

Housing costs are sky-high in Hong Kong. A lot of people find they don't have much left over once they have paid the rent, and even those with decent jobs can find themselves wedged into apartments not much bigger than a parking space. It's no wonder that the Hong Kong government's 'Lantau Tomorrow Vision' – a plan to provide new housing for 1.1 million people, with 70 per cent promised to be public housing – is attractive. East Lantau Metropolis is due to be constructed over the next thirty years on 1700 hectares of new islands.

Lantau is Hong Kong's largest island. It's a hilly oasis of green, home to the 'Buddhist Five Zen Forests', with the flat expanse of Chek Lap Kok International Airport sprouting off its northern side. The 'new vision' is for three islands on its eastern side. The official plan shows that the first to be built will completely surround the small uninhabited island of Kau Yi Chau, which would be retained as a park sitting at the centre of the new townscape. The second phase will be the Hei Ling Chau artificial islands, which will be squeezed between a number of natural islands, turning the landscape into a hopscotch of urban and green islands.

Everyone agrees that the current scarcity of housing is unacceptable. The Hong Kong government has said that if

East Lantau Metropolis

CHEK LAP
KOK AIRPORT

DISNEY
LAND

KAU YI CHAU

LANTAU

forest

forest

SUNSHINE ISLAND

HEI LING CHAU

mountains

mountains

reservoir

KEY   4 km

——————— PLANNED RAIL

━━━━━━━ PLANNED ROAD

+ + + + + POSSIBLE RAIL

- - - - - POSSIBLE ROAD

▰ EAST LANTAU METROPOLIS

source: based on 'Lantau Tomorrow'
(2019)

new homes are not going to be built in open water then
Hong Kong will have to sacrifice some of its much-loved
countryside. Amy Cheung Yi-mei, a director from the government's Planning Department points out: 'If we are able
to develop the East Lantau Metropolis, we would not need
to touch the country parks.' With the new islands, the government argues, a balance between nature and development
can be struck. The islands will be plugged into road, rail
and air transport links, turning Lantau into a major international economic hub. The idea is that, where possible, the
new land will be bulked out with local construction waste

(Hong Kong produces about 1500 tonnes of rubble a year) rather than imported sand. There are other green promises. The new islands will have 'eco-shorelines' to nurture bio-diversity and will be part of a wider programme of nature conservation across Lantau.

The plans for East Lantau Metropolis have met with serious push-back. A decade or two ago this kind of mega-project would have sailed through – just another incredible feat following in the wake of the new airport and the Hong Kong–Macao bridge. But the mood has shifted. When critics point out that the new islands could be underwater by the end of the century, they are listened to. Creating homes for such a huge number of people on a coastal 'frontline' that is set to be pounded by typhoons and swamped by sea-surges seems a curious way of spending $64 billion (this is the estimated cost, though some say it will be double that). The project is also highlighting Hong Kong's fractious relation-ship with mainland China. Its advocates think it will help preserve autonomy but many Hongkongers regard it as just another mainland imposition. It's hard to find anyone in favour of the 'Lantau Tomorrow Vision' in the comment pages of Hong Kong newspapers: 'Like all major HK recent projects, this useless idea will take close to 29 years to mate-rialise. By that time HK will be part of Shenzhen with a lot more available land and more beyond'; 'By the time it's fin-ished, it will seem ridiculous to have spent so much time and so much money on making a small piece of land in the sea.'

For others the main issue is the destruction of the green island of Lantau. A Save Lantau Alliance has sprung up, arguing that 'the protection of Lantau Island is not only to defend Hong Kong's back garden and the last piece of pure land, but also to say "no" to the government's violent rape of public opinion and blindly following the Mainland!' Protesters have taken to the streets, and public opinion, which might once have been assumed to be desperate for new homes, seems to have turned. In part this is because the claim that there is nowhere else to build has been disputed by those who point to large areas of undeveloped and 'brownfield' land in the city, especially in the New Territories area. Tom Yam from Save Lantau Alliance says: 'The government wants to take the path of least resistance: the middle of the sea', even though 'brownfield patches remain untouched'.

In the Gulf States and now in China, island-building has started to become a default option for growing coastal cities looking for more land. Island-building is tried, tested and cost-effective. The Hong Kong government says that the cost of creating land on new islands is up to 30 per cent cheaper than building on brownfield sites and that it makes more sense to have housing close to the city centre than in the relatively distant New Territories. It's also true that properties with sea views still sell for more than those staring over urban sprawl so, whatever the promises of social housing, developers can look forward to healthy profits.

But something is missing: a proper appreciation of the climate changes and disappearing shorelines that threaten Hong Kong. The master plan, which goes under the title 'Preliminary Concepts for the East Lantau Metropolis', gives climate change just one cursory sentence: 'The reclamation level and infrastructure at the coastal areas should be resilient to extreme weather conditions.' It's a throwaway line, an afterthought. East Lantau Metropolis may well be home for 1.1 million people but surprisingly little attention has been given to its future.

Not an Ending

ISLANDS ARE RISING and falling. We keep building islands even as natural islands are disappearing. The new ones are not very high and they are vulnerable to storms and sea-surges. Are we crazy?

It's a serious question. In the HQ of a Dubai developer looking at maps of The World and The Universe – or watching the film of China's Ocean Flower, whose leaf-shaped islands jostle with high-rises and enclose a lotus-shaped island of medieval castles and rollercoasters, or entering the gloom of a hut made of palm fronds in Panama, where each winter the water swirls across the sandy floor – it seems an inevitable question. The delirious, abnormal quality of many artificial islands is brazenly celebrated, and any species that wilfully wrecks its habitat can justifiably be accused of losing the plot.

So are we crazy? I guess we must be. But it's an infectious madness. So many yearnings are being concentrated on islands – these little spaces of escape, delight and fear. And as long as people will pay over the odds for the privilege of looking out over water, island-building will continue to be a moneymaker. The story of relocation will continue too – as long as people in places that are disappearing are unable or unwilling to protect themselves. Relocation is a chapter in this story that has barely begun. We are on its first page, perhaps even just its first line. The rises in sea level seen so far are as nothing to what is predicted. Hundreds of millions live on vulnerable coastlines and every year more arrive.

We can't keep away from the sea even though we know it is dangerous. It's a perilous love affair, a deep need that goes beyond the pursuit of wealth, exclusivity or glamour. The central question of *The Drowned World*, J. G. Ballard's 1962 science-fiction novel about an inundated, overheating planet, is why people don't do anything to stop it. Ballard's curious speculative theory is that an 'atavistic' part of the brain has been triggered – a primal urge to *go back* to the place where we first evolved, and to slip, slither and fall away into the amniotic ocean.

Shorn of the idea that all we ever really wanted is to grow back our gills, and rephrased as an evolutionary predisposition to seek out and be close to water, 'aquaphilia' does help explain something of our current dilemma, of our simultaneous running towards and away from the threatening shoreline. At the start of this book I dragged up a memory: of how, as children, my brother, sister and I would scramble across to a tiny island in a patch of woodland called Wintry Wood and then stand there, wondering what to do next. It's an experience I've had many times on my island travels. For each island I've wanted to get to, I've had to plan my trip very carefully, not having the time or the money for a second attempt. The build-up has been long and costly but, once there – once I'm safely ashore – well, what then? There is not much to do on small islands. It's just like back in Wintry Wood. Elation and restlessness. It was the journey that mattered. The heart of any island is the water crossed.

Artificial islands are older than recorded history but the most ancient examples tend to be modest and low-lying affairs even when they required untold generations to build them. By contrast, many twenty-first-century islands are eye candy, whoopingly unnatural. They are deranged. But with that off-kilter ambition comes hope, for all things seem possible. Time and again, from Phoenix Island to Chek Lap Kok and from Ocean Reef to Fiery Cross Reef, I've been startled and – despite everything, despite knowing better – delighted at the audacity and creativity of our age of islands.

The big question is this: is it possible to harness all that boldness and inventiveness for schemes that are sustainable and of value to a threatened planet? We know that islands can be key to coastal protection, that they can act as flood defences. If built wisely they could provide protection, farmland and living space as well as enhance terrestrial biodiversity and increase forest cover. Flevopolder, the oldest of our era's unnatural islands, is still the biggest and – despite its flat, geometric landscape – in many ways the most impressive.

Not only does it provide arable crops, flood defence for a swathe of the Netherlands, and houses for people of all incomes, but recent years have seen a determined effort both on Flevopolder and on small, wildlife haven islands off its shores to provide habitats for species other than our own.

Building an island always comes with costs but if they are going to be built, we should insist they add more than they subtract from the environment. There are various international protocols already in place that nudge developers in this direction but they clearly aren't working. A robust evaluation would look at the whole operation – from the energy spent in digging the first spadefuls of sand to the ongoing resources allocated in keeping the lights on and the toilets flushing in 'ultra-star' hotels – and insist that, overall, there is a net benefit to climate change and biodiversity. That might drain the blood from some developers' faces but, as we have seen, the watchwords of modern island-building are boldness and ambition; the impossible is being built in such places every day. That unstoppable energy is just what we need, but channelled towards a greener, long-term future.

Thinking up new islands is a lot of fun. I couldn't leave this journey without coming up with one of my own. Here it is, set in the wide bay in eastern England called The Wash. I've often thought it was odd that, almost within foghorn distance of the industrious Dutch, this large, square-shaped and shallow inlet has been left fallow. It does have a couple of small infrastructure islands, built and then abandoned in the 1970s to test the feasibility of creating a tidal barrage, and much of the surrounding land was reclaimed from the sea centuries ago. Today, a significant portion of eastern England is threatened with flooding. A new island in The Wash wouldn't solve that problem but it could make any inundations more manageable and less disastrous. Guthlac Island is named after St Guthlac, a local island-dwelling hermit who was once much venerated in the region. It would be a wetland, some 1521 square kilometres in size, given over to marsh and river species now extinct or largely diminished elsewhere in the UK. It would be dotted with small

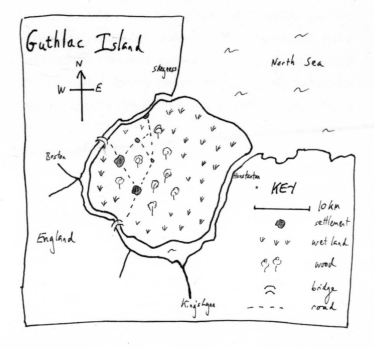

settlements and, to make it financially possible, a couple of towns. It would be spectacular but nothing like Dubai; Guthlac would offer a different type of amazement. It would be an island for geese, herons, otters and eels as much as for people – one designed for today and for tomorrow.

I started this book with a pull of the oars that took me ashore an unnamed crannog, one of many antique artificial islands that sprinkle the lochs of Ireland and Scotland. We've been building islands for a long, long time. They have been disappearing for a long time too. Those two things have now speeded up and changed in scale and meaning. Island-building is not working with the planet anymore; it has been malformed by hubris and greed. The remnants of much older human-made islands are still there, signposts to the brevity of our lives and our era. Unvisited and unnamed like the crannog, little more than a black lump of lifeless stones in a cold Scottish loch. Yet, for all that, it catches the eye and the prow draws closer.

Acknowledgements

Many people have helped me with this book over the past few years, given me their time, shown me hospitality or simply shown me the way. My editors – James Nightingale, Charlotte Atyeo and Mary Laur – have also provided invaluable assistance. Special thanks are due to my Rachel, Louis and Aphra and also my mum, Shirley Bonnett, for diligently reading through the manuscript and trying to sort out my punctuation.

Bibliography

Appleton, Jay, *The Experience of Landscape*, John Wiley: London, 1975

Baldacchino, Godfrey, *A World of Islands: An Island Studies Reader*, Institute of Island Studies, University of Prince Edward Island: Prince Edward Island, 2007

Ballard, J. G., *The Drowned World*, Victor Gollancz: London, 1962

Bauhaus, Eric, *The Panama Cruising Guide*, Sailors Publications: Panama, 2014

Cronin, William, *The Disappearing Islands of the Chesapeake*, Johns Hopkins University Press: Baltimore, 2005

Díaz del Castillo, Bernal, *The True History of the Conquest of New Spain: Volume One*, Routledge: Abingdon, 2016

Displacement Solutions, *One Step at a Time: The Relocation Process of the Gardi Sugdub Community in Gunayala, Panama: Mission Report*, Displacement Solutions, 2015, accessed at: http://displacementsolutions.org/new-report-on-the-planned-relocation-of-the-gardi-sugdub-community-in-gunayala-panama/

Gutiérrez, Gerardo, 'Mexico-Tenochtitlan: origin and transformations of the last Mesoamerican imperial city', in

Yoffee, Norman (Editor), *The Cambridge World History: Volume 3, Early Cities in Comparative Perspective, 4000 BCE-12000 CE*, Cambridge University Press: Cambridge, 2015

Herodotus, *Histories*, Hackett: Indianapolis, 2014

Hong Kong Government, *Hong Kong 2030: Preliminary Concepts for the East Lantau Metropolis*, Development Bureau and Planning Department: Hong Kong, 2016

Howe, James, *A People Who Would Not Kneel: Panama, the United States and the San Blas Kuna*, Smithsonian Institution: Washington, D.C, 1998

Lawrence, D. H., 'The Man Who Loved Islands' in *D. H. Lawrence: Selected Stories*, Penguin: London, 2007

More, Thomas, *Utopia*, Cambridge University Press: Cambridge, 2016

Quirk, Joe, *Seasteading: How Floating Nations Will Restore the Environment, Enrich the Poor, Cure the Sick, and Liberate Humanity from Politicians*, Free Press: New York, 2017

Records of the Grand Historian: Records of the Grand Historian of China: The Age of Emperor Wu, 140 to circa 100 B.C., Columbia University Press: New York, 1961

The Rough Guide to Dubai, Rough Guides Ltd: London, 2016

The Seasteading Institute, *The Floating City Project: Research Conducted between March 2013 and March 2014*, The Seasteading Institute, accessed at: http://www.seasteading.org/wp-content/uploads/2015/12/Floating-City-Project-Report-4_25_2014.pdf

Thomas, Charles, *Exploration of a Drowned Landscape: Archaeology and History of the Isles of Scilly*, B.T. Batsford: London, 1985

Tuan, Yi-Fu, *Topophilia: A Study of Environmental Perception, Attitudes, and Values*, Prentice Hall: Englewood Cliffs, N.J., 1974

Index